C000281789

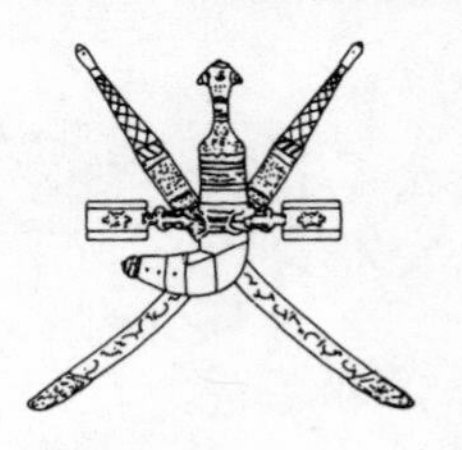

THE SEA OF SANDS AND MISTS

The Story of the
Royal Geographical Society
Oman Wahiba Sands Project
1985–1987

Patron:
His Royal Highness Prince Michael of Kent

Corporate Patrons:
Assarain Enterprise
Gulf Air
Land Rover Ltd
Mohsin Haider Darwish
Racal Electronics
Suhail and Saud Bahwan
Taylor Woodrow-Towell
Zubair Enterprises

Brigadier J. T. W. Landon

THE SEA OF SANDS AND MISTS

Desertification: Seeking Solutions in the Wahiba Sands

Nigel Winser

CENTURY

LONDON SYDNEY AUCKLAND JOHANNESBURG

First published in 1989 by Century Hutchinson Ltd,
Brookmount House, 62–65 Chandos Place, Covent Garden,
London WC2N 4NW

Century Hutchinson Australia Pty Ltd, 89–91 Albion Street,
Surry Hills, Sydney, New South Wales, 2010, Australia

Century Hutchinson New Zealand Limited, PO Box 40–086,
Glenfield, Auckland 10, New Zealand

Century Hutchinson South Africa (Pty) Ltd, PO Box 337,
Bergvlei, 2012 South Africa

Printed and bound in Great Britain by
Butler and Tanner Ltd, Frome and London

British Library Cataloguing in Publication Data

Winser, Nigel
The sea of sands and mists: desertification – seeking solutions in the Wahiba Sands.
1. Oman. Deserts. Exploration
I. Title
915.3′53040453

ISBN 0–7126–1609–8

Contents

For Shane.
For all People of Sands.

Illustrations

Page 1

A good clear view of the Wahiba Sands as seen from a satellite on the 13 March 1986. This Arabian sand sea desert, approximately the same size as Wales and home for over 3,000 nomadic Bedu, is one of the most remarkable sand deserts in the world. Situated in the north and east of the Sultanate of Oman it has recently become the focus for long-term desert research. No desert in the world has such a broad range of sands and dunes so neatly compressed into such a precise area. This isolated sand sea is bound by the Wadi Batha to the north and east, the Arabian Sea to the south east and Wadis Matam and Andam to the west. The date-palm gardens of Mintirib can be seen at the very top of the Sands and woodlands to the east are seen as dark patches. By comparison to other sand deserts, here is a complete and pristine desert ecosystem providing a perfect laboratory for earth, life and human scientists wishing to add to the knowledge of sands seas of the world. (*Image processed by Chris McBean, Durham University, for the Royal Geographical Society Oman Wahiba Sands Project*)

Pages 2–3

TOP, EXTREME LEFT: Most mornings, mists could be seen in the valleys between the dunes and in the wadis, quickly disappearing as soon as the sun appears. Here the last traces in the Wadi Batha can be seen early one morning from the top of the high dunes near Mintirib. (*Michael Keating*)

TOP, LEFT: Throughout the Sands there are isolated prosopis trees that tap the groundwater with their deep roots as well as absorb water from the night mists through their very special leaves. For the team, many such trees were useful signposts when travelling along the routes used by the Bedu. (*Richard Turpin*)

TOP, RIGHT: Three project Land Rovers finding a route through the large dunes by the coast, bypassing an outcrop of aeolianite. Considerable time was spent reconnoitering the Sands both from the air and on the ground, made possible by the logistic support provided by the Sultan of Oman's Armed Forces and in particular a small patrol unit, the Coast Security Force. All vehicles were in constant radio communication with Taylorbase and always travelled in pairs when leaving the main routes. (*Richard Turpin*)

TOP, EXTREME RIGHT: 'So long as the sun and the moon continue to rise – Oman and Britain will remain friends.' Here the Almighty keeps his side of the bargain at our base in the prosopis woodlands on the east side of the Sands. (*Richard Turpin*)

BOTTOM, LEFT: Taylorbase – home for the project and centre of all field operations situated near Mintirib on the edge of the Wadi Batha. This 'field university' was the vision of John Cox and Terry Tobin of Taylor Woodrow-Towell offering full laboratory, computer and conference facilities, providing good opportunities for those interested in the Sands to meet and discuss matters with our team of 35 scientists during our four-month stay. This was taken on the 1 April 1986, the day we departed and the camp dismantled. (*Richard Turpin*)

BOTTOM, RIGHT: To properly understand an area, geographers today collect information from a broad spectrum of sources. Harnessing the best of new technologies, particularly satellites and computers at one end, is balanced by the invaluable information held by the guardians of a region and passed down over thousands of years. Here Said Jabber Hilays Al Wahibi (right) discusses the effects of wind on sand movement, with Nigel Winser, Roger Webster and Andrew Warren. (*Richard Turpin*)

Page 4

TOP, LEFT: The camels of the Wahiba Sands are noted for their speed and stamina, as we were to find during a crossing of the Sands, following Wilfred Thesiger's original route. Arabist Roger Webster was called affectionately 'Dhiyab' (the Wolf) by the Bedu, with whom he and his team spent many happy days. (*Richard Turpin*)

TOP, CENTRE: Eighty-seven species of birds were recorded during the project and many of these were attracted to the prosopis woodlands. Michael Gallagher from the Natural History Museum in Muscat was kept busy with his mist nets, here seen holding a Bruce's Scops owl. Author of *Birds of Oman* and an experienced field biologist, we were fortunate to have him 'on loan' from the Oman Government. (*Richard Turpin*)

TOP, RIGHT: Not all foxes in the Sands have radio-aerials yet. Ian Linn's study of the larger mammals of the Sands led to a number of new finds. Here he holds a Ruppell's fox which he has just 'radio-collared' to help him understand the range and territories of the highly adapted desert mammal. Having studied the behaviour of mammals all his life it is not surprising he has an uncanny knack of knowing what is going on at night from tracks and scats. (*William Buttiker*)

MIDDLE, LEFT: Throughout the Sands, the Wahiba Bedu offered warm hospitality and friendship to all who passed by their little palm-leaved camps. Here a family display some of the camel trappings that are such a feature of the Sands. The craft of weaving by the women has a long history in the Sands and sadly we noted that it was now on the decline, despite it maintaining high standards and fetching good prices in the markets. (*Richard Turpin*)

MIDDLE, CENTRE: Collecting pieces of the geographical jigsaw – sifting the sands with a kitchen sieve was one of the many methods Willie and Sonya Buttiker used to determine the vast range of animal life that lived in the Sands. Here dew-drinking beetles are being investigated – the first evidence of these in Arabia and similar to those found in the Namib Desert. It must be good fun studying invertebrates as every day brought new surprises. By the end of the project over 16,000 specimens had been collected and despatched to over 60 specialists for identification all over the world. (*Richard Turpin*)

MIDDLE, RIGHT: The long-term task by scientists to check data provided by satellite imagery is enormous. Detailed studies of smaller areas can help 'ground-truth' information, here being done by Simon Kay with the help of a Magnavox navigator linked to orbiting satellites and colour aerial photography so recent that it even included Taylorbase. Each piece of information is added to a computer programme and given a spatial code which in turn can be correlated to the different colour spectral signatures or pixels (as they are known) that comprise satellite imagery. These 'geographic information systems' will become a powerful tool for all conservation and development projects. (*Richard Turpin*)

BOTTOM, LEFT: The sheer size of many of the trees in the Sands surprised us all. The importance of the *Prosopis cineraria* woodlands in and around the Sands cannot be underestimated. With trees growing up to 20 metres high it is not surprising they provide a home for a rich variety of flora and fauna as well as shade and building materials for the Bedu. The tree's ability to withstand tremendous drought and survive by absorbing the night mists make it a suitable contender for the tree that may one day be used to halt other encroaching deserts. Kevin Brown adopted a number of techniques to reach parts not yet eaten by the goats and camels. (*Michael Keating*)

BOTTOM, RIGHT: Where sand and sea meet. Divers of the Sultan of Oman's Navy assist Bob Allison in his task to decide whether the sands had originally come down from the mountains or up from the sea. With samples from over 400 locations, Bob has designed a model of how the sands might have formed. The rock outcrop which Bob is using to hang his towel is aeolianite, the hardened remains of an earlier sand sea which has since become hardened over time. Rita Gardner surveyed this and found it to be the most extensive example of aeolianite found anywhere in the world (*Richard Turpin*)

Page 5
The two large prosopis trees at Tawi Haryan offered valuable shade to Said Jabber, Dhiyab and the author during their journey by camel across the Sands. We rested here during the heat of the day, while the camels grazed and drank from the well and Said cooked us a meal. Little has changed since Wilfred Thesiger passed this way some forty years earlier on his journey northwards. For a brief period Dhiyab and I had the privilege of sharing with Said a pace of life now lost. Here and only here, did we really begin to understand the sheer enormity of the Sands. (*Nigel Winser*)

Page 6
TOP: His Royal Highness Prince Michael of Kent, ever active as the project's Patron, joined the team in the field on two occasions. Here Said Jabber Hilays and his family entertain the Prince and the British Ambassador Robert Alston (left) following a busy morning learning the art of camel riding. Dhiyab is seated on the far right and unusually is not dressed in his *dishdasha*. (*Nigel Winser*)

BOTTOM: Said bin Jabber bin Hilays Al Wahibi and family in the Sands. (*Simon Ferrey*)

Page 7

TOP: Touching the past. Said, Dhiyab and the author enjoy coffee in the very early morning before setting off for a long day across the Sands. Few travel by camel today, so there was much talk of the old days before the pick-up truck. We had spent the night with the Bedu who were living on the west margin of the Sands, where the rangeland was particularly lush following some recent rains. (*Nigel Winser*)

BOTTOM: The guardians of the Sands. At the end of the project, Sheikh Mohammed bin Hamad bin Daghmal Al Wahibi, the Head of the Wahibah people hosted an enormous feast of 'several goats' for the project team at his house in Al Falaj. (*Simon Ferrey*)

Page 8

TOP: Mintirib hiding behind the high mega-dunes at the north end of the Wahiba Sands, which have been kept in check by the regular flow of the Wadi Batha. It did so during our visit and the raging torrent was over 6 metres deep. This photograph was taken soon after the rains and the views to the Hajar Mountains in the north are unusually clear. (*Nigel Winser*)

MIDDLE: A familiar sight for those who stayed at Taylorbase. Looking towards the north wall of the Sands, with the Wadi Batha in the foreground, here covered in an early morning mist. (*Michael Keating*)

BOTTOM: The greening of deserts. Deep in the woodland community of the prosopis trees, a few days after a period of rain. Is this species the tree that can assist those seeking solutions to desertification? (*Nigel Winser*)

Foreword

Having always had a great fascination for geographical expeditions I was thrilled and proud to be asked to be Patron of the Oman Wahiba Sands Project. I had the opportunity on two occasions to join the expedition in the Sands, and was impressed by the high standards of professionalism and teamwork, not to say enthusiasm, of its members. One of the things that struck me was that however arcane some of the scientific experiments that were being conducted appeared to be, they all fitted neatly into the overall strategic plan.

The hospitality and friendship shown to me by the people of the Sands was unforgettable. My links with them were strengthened further when I was presented with a splendid camel by the project guide, Said Jabber. Riding it proved a not altogether enjoyable process, but the kindness and generosity of Said was typical of the Bedu I met.

The Sea of Sands and Mists brings back many memories of the beauty of the Sands, their mystery and constantly changing colours and the surprising richness and variety of life. Nigel Winser is an inspiring and tireless expedition leader. Now he has written this compelling book so that others can share his experiences. I have immensely enjoyed reading it. I know you will too.

Michael

PRINCE MICHAEL OF KENT

Introduction

The stiff swing doors at the entrance of the Royal Geographical Society have long been busy with those wishing to push forward the frontiers of geographical knowledge. They are still fairly stiff, sixteen years after I nipped my fingers on the first occasion I passed from the bustle of Kensington Gore into the hushed and hallowed entrance hall of the Society. The presence of those successful travellers who have returned to offer explanations and descriptions is almost palpable.

'Meet me at the RGS at six', I had been instructed by the President. It was 1800 hours precisely. The commanding voice of Sir Duncan Cumming could be heard clearly as he strode down the main hall to welcome me to the Society. We had met three weeks earlier when he came to speak to our college exploration society and had invited me to attend one of the Society's lectures, to meet some of the staff and Fellows. I can remember this first introduction to the Society clearly: it opened up a world of hopes, visions and important challenges. Much was placed on my personal agenda. I sensed energy and urgency.

The connections were powerful. Meeting active travellers who still had snow, sand, mangrove swamp mud on their boots was exciting enough; to hear of their priorities at first hand was a stimulating privilege. To see the marks of well-known explorers in the library, in the map room and listed on the walls of the lecture hall was simply inspirational. Here was the centre of 'ancien regime' exploration, now moulded by a new breed of scientific explorer with contemporary technological skills.

I wanted to explore. Cumming blooded me that day. Appropriate for Kensington Gore. To have decided before the end of the evening to join the Society, to meet the Director, John Hemming, and for a new chapter to have opened, all within a space of a few hours, is the stuff of the RGS. The Society is a meeting place, indeed a sort of geographical *suuq*.

Although Dr Livingstone and Ernest Shackleton honour two platforms outside the RGS building, once inside this red-brick Norman Shaw, there is no prejudice for a particular environment. Prospective polar explorers are treated with the same respect and service as those wishing to study coral reefs in the Indian Ocean or mangrove forests in western Australia. That is just as well as there is much to be done. Planet earth is still a scientific wilderness. Too many pieces of the geographical jigsaw have yet to be found let alone be strewn on the environmental carpet for decision-makers to try to order. And of course the burden lies with a new generation of geographers who are only just shouldering their responsibilities. As contemporary caretakers, our reputation at the beginning of the next millennium is at stake.

Geography covers a broad spectrum of interests and disciplines and its members boast a long history of research and concern about the world. But as the pressures of overpopulation mirror those natural whims of the Almighty, is it not the geographer who has the responsibility to monitor, predict and draw the boundaries of development strategies throughout the world? But why are there so few geographers in Parliament or politics generally? What influence does this discipline have? Caught between the Royal Society and the British Academy, geography has always been 'twixt the two. Art or science? This dichotomy is surely its very strength. Geography breaks down barriers between the physical and life sciences; it encompasses the social sphere as well as the better promoted bio- and geo-spheres. Geographers do agree they must provide leadership through proper explanation in the future. And Lowther Lodge is one of the more convenient meeting places in Europe for discourse on all matters geographical.

The Royal Geographical Society has been the British home for the organization of expeditions overseas for the relatively short time of 158 years. Chartered to 'further the knowledge of geography, that most entertaining branch of science', it has attracted scientists, politicians, businessmen, overseas visitors and interested travellers from the very beginning. The exploring tradition is still strong today. A short walk in the basement of the Society among the map cabinets and through the archive library overlooked by pictures of expeditioners past and present, a chance to peek into the many drawers housing thousands of maps, touches the very stuff of exploration. Livingstone's cap is almost on the same shelf as a satellite navigator, or a field computer. Photographs range from a dusty and slightly faded print of the first siting of an obscure and yet to be named waterfall (1886), to a team using field computers to study sand movements in an Arabian desert (1986). That innate curiosity, known to explorers as the 'Ulysses factor', is easily stirred. This powerful spirit has taken teams to the top of Everest and to the bottom of the oceans. Fortunately we all still have this powerful urge to look over the top of a distant hill, and fortunately the Society is very much in touch with many who now harness it to venture overseas with scientific notebook in hand.

And so, bitten by the Society and not just its doors, I am still here. Since joining the Society as a volunteer in 1976, hotfoot from undergraduate expeditions following the course of the Tana River in Kenya, the forests of West Ethiopia and the sands, insects and rock art of Central Sahara, I have had the privilege to be in the front line of three very different scientific expeditions: fifteen months in the rain forests in Sarawak, East Malaysia (1977–8), six months in the high Karakoram mountains and glaciers in Pakistan (1980) and four months in the Acacia/Commiphora scrublands of the Kora National Reserve in Kenya (1983). These have offered valuable experience in looking after scientific endeavour in distant locations and under the leadership of three characters, each with their own methods of achieving high scientific goals; my apprenticeship has been colourful and rewarding. Expeditions produce more than the sum of the parts. They release tremendous energy sometimes called 'eco-adrenalin'. Fuelled by individual commitment, it is no wonder that so much work can be done in so short a time. Over 600 scientific papers testify to the energies of a total of 230 earth, life and human scientists with whom I have had the pleasure to work over the last decade. Here that innate curiosity is being harnessed for scientific endeavour. Multiply this by the whole exploration work-force and we have a corpus of many thousands of green explorers who collect geographical information about the world; 400 expeditions are on file each year, made up of schools, universities, adventure and sporting clubs, youth organizations, environmental research groups, aid agencies and many more field teams with ages ranging from 17 to 70. Their results, although not as startling as those who have been before, are no less valuable. Their enthusiasm and energy is recognized by the Society and considerable effort is spent to help train and advise this growing body of information-gathering explorers. Co-ordinating this is the responsibility of the Expedition Advisory Centre, one of the many offices that are housed at No 1, Kensington Gore.

Exploration need not just be for the individual who wants to stand atop a new mountain, but in some ways more importantly, for those who want to contribute geographical knowledge for development and for conservation. Keeping a hand on the pulse of the expedition game is very much my responsibility in the Society's Expedition Office.

But there are changes to be made. Collecting information about plants, animals and their environment is not enough. Pure academic research must address applied needs. The plight of the nomadic pastoralists in

Africa who have to cope with wood-fuel shortages and floods is critical. But the felling of forests is now widespread and continues to reduce the overall biomass of the land, causing irreversible biological and geomorphological problems. Shake the network too hard and permanent rips will appear. Next year the situation will be worse. The year after more so. Geographers might have been able to predict this particular problem, but they failed to convince government decision-makers to take the necessary steps before it was too late. Nor can geographers immediately offer a solution for arid areas. Why have we not undertaken enough research to identify a fuel-wood tree that could be planted to avoid these pressures over the next twenty or thirty years?

So there is an urgency for geographical information. For the first time in man's history, he can alter his environment. Patience can no longer be our cry. For instance, why have we not yet managed to convince heads of state that the economic resources of forests and woodlands are more valuable than oil, gold and diamonds? The simple answer lies with our comparative lack of knowledge and skill at communication. Geographers have to become much better at being able to provide information to decision-makers in a language that can be clearly understood. Never before has there been a need for politicians, businessmen and decision-makers to hold hands with the information-gatherers. In turn those who collect the data must be better communicators, both to those who are expected to use it and also those who are the guardians of the areas concerned. Close liaison with the local community, accounting for their own needs and priorities, has long been the failure of the large development programme schemes. Making sure that environment exploration is channelled to help the tribal landlords, on whose land we all have trespassed in the past, is the responsibility that the RGS Expeditions Committee is today taking very seriously.

Detailed geographical knowledge of any community or ecosystem must be the basis for wise management and development throughout the world. The European large blue is as much an indicator of remaining natural resources as the large tracts of tropic desert land now barren of farming communities, wildlife and plants. There is much to be done. Those scientific papers are a start, but they barely scratch the surface and they do not necessarily reach the front line decision-makers in government who handle day-to-day management problems. Today's geographers are going to have to do better. The Society must continue to identify geographical priorities and so provide a forum for international meetings and discussions on a broad range of interested disciplines.

The selection of locations for Society expeditions is delegated to a committee, which calls on advisers to identify priorities. They meet round the 25-foot-long oak refectory table in the Council Room. With only one project every two to three years, each environment has to take its turn. This can lead to considerable discussion. Asian rain forests, Himalayan mountains and glaciers, African bushland and the magnificent Kimberley wilderness in north-west Australia have all recently hosted an RGS team. Indian Ocean marine research, Pacific tropical islands, Chinese caves, relic African forests, South American moist forests, Central Sahara Ahnet, the Egyptian Quattara Depression and the Bramaputra valley have all been mooted as important locations. Spokesmen have shared their ambitions across the Council Room table. Competition between these ambassadors is of course amicable, but the desire for one's chosen region or discipline to be adequately represented can lead to periods of academic scurrility. All grist to the mill.

In 1983 the desert scientists were beginning to lobby for a project to one of the world's less-known environments – the sand deserts. The RGS was just embarking on a research programme in the relatively arid scrublands of the Kora National Reserve on the banks of the Tana River in Kenya. This was the home of George Adamson who was still busy rehabilitating lions back to the wild. Deserts have long been a fascination for Britons, and there was no shortage of experienced arid-zone researchers who supported the proposal. Committees can be notoriously slow at deciding where to go; political, economic and scientific priorities have to be carefully considered. There are over twenty sand deserts of the world and each could provide a suitable research location.

Our general knowledge of arid areas of the world is notoriously weak. More than one-third of the earth's land surface is arid and supports nearly 20 per cent of the world's population. Both of these are expanding, and current misuse of the biological resources is causing considerable hardship. International research has had problems in getting to grips with the workings of desert areas as separate ecosystems. So much of the published material is disparate and unrelated. It is not surprising that governments are confused about long-term programmes. Immediate drought crises and emergency aid programmes cloud still further the priorities. The Society wanted to address this problem. It began to research in very much more detail which desert would be suitable for a detailed in-depth study by an international, interdisciplinary team which could offer an approach through the earth, life and human components of a desert ecosystem. In painting a picture, the artist has to assess resources before deciding the size of the frame. Likewise with desert research. It is more important to cover the whole of a small canvas than merely to touch isolated corners of a larger one.

Sand, usually of grain size between 0.1 and 2mm, is the traditional component of deserts. But only a third of arid lands comprise sandy areas and when they do they move together, either by water or by wind, to form bodies called 'ergs' (the Arabian name for desert with dunes) or sand seas. Most of the world's sand is found in these sand seas, which vary in size between 100 sq km and 560,000 sq km. Sand seas include both active (mobile) and fixed dunes, and the best known are the Atacama (South America), Namib (South West Africa), Great Sandy (Australia), Gobi, Sahara and the great Rub' Al Khali of Arabia. Sand deserts are particularly arid and have had as little rain as 0.4mm in a year. In spite of this, they do not follow the description offered by many world atlases as having no vegetation. Furthermore many people of the world have decided to make sand deserts their home, have become masters of that environment and are quite happy to go on living there.

But how much do we really know about desert ecosystems, let alone the sand components? While desert research bibliographies might be full of published reports, there are few that have managed to study them as complete units. Fewer still have managed to harness the skills of different disciplines to offer a balanced picture. Earth scientists who study how the earth's land surface is formed and the processes (heat, wind, rain, tectonics, cold, etc) that change it, called geomorphologists, have been most active over the past sixty years, but even their work could be described as thin. Desert chiefs Ron Cooke and Andrew Warren offer excuses for this in their now standard reference *Geomorphology in Deserts*:

> Geomorphological descriptions of deserts have tended to be superficial for several reasons. Firstly geomorphology has rarely been the exclusive concern of exploratory expeditions. On many desert journeys, there is competition between the concern for survival in an insecure and potentially hazardous environment, and the collection of scientific information, usually to the detriment of the latter. Secondly where an expedition does not include a geomorphologist, the land forms may be described in an amateurish fashion. Thirdly, because exploratory expeditions are usually concerned with areas about which very little scientific information of any sort is available, all manner of environmental observations may be recorded so that the geomorphological information may be incidental, and the results may be published as an unpalatable farrago of descriptive data. Fourthly although many resource surveys included sections on geomorphology, other considerations are commonly more prominent. A fifth reason is that many expeditions and surveys cover large areas quickly and observations tend only to be of a reconnaissance nature.

Certainly the RGS did not want to add to this 'farrago' and nor did it want to leave out the life and human components. But studies of sand-sea areas are even rarer, largely due to inaccessibility for both physical and political reasons. They need a base camp; an expedition which can provide good laboratory facilities in the field and a wealth of support to enable scientists to work safely and without concern from six in the morning to midnight was something the RGS was keen to offer, especially if it could involve scientists from the host country and accommodate interested parties from the government.

Furthermore the image and value of overseas research was an issue of which the Society was very much aware. Scientific arrogance creates barriers and a sensitivity to development issues. Our approaches to governments are designed to seek a working partnership, and I think it is fair to say that many lessons have been learnt over the last ten years. Per Lindblom at a World Commission for the Environment and Development meeting ephasizes the gap between science and development: 'The problems today do not come with a tag marked energy or economy or CO^2 or desertification, nor with a label indicating a country or a region. The problems are multi-disciplinary and transnational or global. The problems are not primarily scientific and technological. In science we have the knowledge and in technology the tools. The problems are basically political, economic and cultural.' For us this emphasized the importance of breaking down barriers in the field and encouraging international scientists to share a soil auger and discuss the results over a beer in the evening.

There was delight in the Expedition Office in February 1984 when Dr Colin Bertram mooted the idea that the Society should consider mounting a project to the Sultanate of Oman, the only Arabian country to own a small sand sea, easy of access, suitable for research, perfect as an erg and unique in its composition. Colin had just returned from staying with his old fishing partner, His Excellency Said Thuwainy, with whom he undertook the first fish survey along the coasts of Oman in 1948. Colin is no spring chicken. He has been at the centre of geographical exploration for many years now. His fishing report still stands the test of time. Colin, together with his wife, has since undertaken fieldwork in every continent in the world and is a senior Fellow of the Society. Said Thuwainy had been treating Colin to some traditional Oman hospitality. They say that Oman and Britain will be close friends until either the Sun or Moon fails to rise one day. But friendships born and bred on expeditions will always stand the test of time. Colin came back with glowing stories about the changes in Oman and positive impressions concerning environmental research. He was to be the catalyst that helped the RGS to decide on this venue for the next project as his proposal arrived concurrently with that of the desert chiefs headed by David Hall, Ron Cooke and Denys Brunsden, who had similarly pointed a finger at Wahiba, that 'little known yet isolated sand sea' – top of the list of arid areas to visit.

A note from the Director appeared on my desk. 'I have been in touch with Professor Clark at Durham University. He has suggested we make contact with Dr Roderic Dutton from Durham's Centre of Overseas Research and Development (CORD) who has many years' experience working in Oman.'

So began the Oman Wahiba Sands Project 1985–7. I rang Dr Roderic Dutton. 'Can you meet me at the RGS at six?', I enquired.

CHAPTER ONE

Sand of a Thousand Colours

I had to admit I was in some pain under the *Calligonum* bush. I was going to keep that to myself for the time being, though. The sun was directly overhead; there was barely any evidence of shade from the camel stick that was stuck in the sand nearby. I stirred, trying not to disturb my travelling companions. Both were sleeping fitfully and had covered their heads with their *masarra* (headscarves) to keep out the sun and the ants. The camels were now lying down and chewing the cud with their eyes closed.

We had been travelling since before dawn. This bush, a mass of red fruiting flowers brought on by the recent rains, was the largest we had seen and the only offer of shade we were going to get that day. Our canvas water bags had leaked and there was still a very long way to go before nightfall – but desert travellers are no match for the midday sun, particularly apprentices and those unused to camels and their saddles. Said Jabber and Dhiyab had spotted the bush long before I did and we had headed straight for it up a steep sand dune rising some 60 metres above the swale. Said had draped a blanket over the bush to give us a small patch of shade and we were now huddled together trying to avoid the sun. We didn't have enough water to cook up any rice for lunch and had to make do with dates and coffee. This was fine for Dhiyab and me, but Said explained he wouldn't be able to sleep. However, the snoring sounds soon coming from under his *masarra* belied this.

I closed my eyes and began to doze off again. We were in the middle of the Sands, we were thirsty, we were hot, we had a long way to go and the blisters I had collected were beginning to hurt; but we were touching the heart of desert life and it was the most significant journey I was to make in the Sands. Just before I fell asleep I wondered how my family were coping with the blizzards in England.

Suddenly Said was shaking my arm. The sun was now low in the sky and it was time to remount and move on a further 20 kilometres to find a Bedu camp for the night. My blisters had caked and were stuck to my baggy trousers. The coffee pot was on the small fire made from the twigs and leaves we had collected earlier. Dhiyab was already saddling his camel. Said asked if I was OK and I tried to put on a brave front in the face of his amused gaze.

I can vividly remember meeting Said bin Jabber bin Hilays Al Wahibi on his home ground for the first time in January 1985; it was the day I was introduced to the Al Wahibah people. He was with a group of about eight Bedu, sporting *khanjars*, the traditional Omani daggers, camel sticks and snow white *dishdashas* (robes), all eyeing me intently in the new office of the Wali of Mudhaybi, the senior government representative for the area. Said was the tallest of those present, the most aristocratic in appearance, and his eyes sparkled with humour. He knew we would be seeing a lot more of each other in the future.

I had driven down from Muscat, turning off the main road and following a dusty and corrugated gravel road through a couple of villages with green date palms and cool irrigation channels called *falajs* where children were splashing each other. I was late and I could not find the Wali's office. A kind old man pointed me in the direction of the *suuq*, thinking I was after some silver. However, I was soon taking my shoes off outside the front door of a magnificent new building before being ushered into a room with plain white walls, where the Wali was working behind an enormous desk.

He was a powerfully built man, with strong features. 'As salaam 'alaikum.' 'Alaikum as salaam.' After we had exchanged a traditional greeting he invited me to sit down and called for *halwa* (sweetmeats), *tamar* (dates) and *qahwa* (coffee). We took stock of each other. The Wali's *masarra* was pristine white, as was his *dishdasha*; the government-issue *khanjar* was clearly solid silver and had a magnificent horn handle. There was no mistaking who he was, and everyone in the building hurried to do his bidding. The letter of introduction from Sheikh Ahmed bin Saif Al Mahrouki at the Ministry of the Interior, forewarning the Wali of my visit, had arrived that morning and now lay on top of the pile of documents on an otherwise clear desk.

This was the fifth Wali I had seen in two days and one of the most important, for his jurisdiction covered much of the western region of the Sands and was nearest to the centre of the Al Wahibah lands. His blessing and support were crucial. Furthermore it was he who would be appointing the local Bedu who would act as our advisers, guides and informants. One of the major marketing outlets for the Sands was the nearby town of Sanaw; situated on the main trading route down the west edge of the Sands by which many of the Bedu buy and sell their produce, it is a rapidly expanding town which has become a focus for the sale of fish, goats, camels, other livestock such as cocks and hens, dates, fruit and vegetables. It was an important market for us to visit to learn about the agricultural activities of the Bedu so it was just as well that our intentions would be understood from the beginning.

I was now familiar with the practice of dipping one's fingers into the sweet *halwa*, having received some of the best *halwa* in the land at the other Wali's offices. The Wali pointed out with some pride that this was from Nizwa, whence comes the crème de la crème of *halwa*.

Qahwa was served. We had three cups each, then rocked the cups to signify we had had sufficient. The tables cleared, we got down to business. I could see Said Jabber settling down, now squatting on the floor with the other Bedu. They were a formidable sight, their appearance living up to their reputation of being a fearsome tribe with a long list of battle honours. Being a newcomer to Arabia, I still had slight misgivings about the possible course of action should I commit even the smallest faux pas, but I needn't have worried.

The meeting lasted an hour and I tried to explain why a team of Britons wanted to visit this particular corner of Arabia and described the precise area we wanted to work in. The Wali replied that he had been fully briefed by the Ministry of the Interior and that he and his people would do all they could to make our stay comfortable and welcome. He told me with whom I should make contact, in particular the chief Sheikh of the Al Wahibah people, Sheikh Mohammed bin Hamad bin Daghmal Al Wahibi, who lived near Aflaj, about an hour's drive from here.

Oil dominated the discussion for a while. Was there oil to be found in the Wahiba Sands? I explained that we were not connected to a mining company, although we were 'mining' facts about the animals and plants with the view that the information to be gained would in fact be more valuable than oil in the long term. In reply to his question about which university I was from I explained that although my initial training was as a zoologist, I was now a *mudiir* – a manager of scientists and administrators. We both laughed and the Wali made a joke about paperwork and red tape. I nodded, appreciative that we had an understanding.

The Wali then turned to the eight Bedu who had been waiting patiently. There was a slight buzz of anticipation. He asked me how many guides I required and, when I replied that for the time being I needed only two, he picked out Said bin Jabber bin Hilays Al Wahibi and a younger-looking and smartly dressed Bedu, Sheikh Sultan Nasser Mansur Al Bu Ghufayli.

Said and Sultan were then given a quick briefing. They both nodded furiously. We bid our farewells to the Wali, who made a parting comment to Said about my being improperly dressed.

Outside in the courtyard, Said and Sultan looked at me more intently. They were aware that they had just joined the team and the implications for them and their families over the next two years began to register. I asked what the Wali had meant about my being improperly dressed. Both roared with laughter. Said handed me his camel stick and Sultan pulled out an expensive-looking headscarf, the ubiquitous *masarra*, and tied it to my head. They both said that I looked *mumtaz* (magnificent) and that if I was to be a true leader of men I should never be seen without either.

We made an unusual sight for those who saw us speeding down the road on our way to Mintirib, for we were like three schoolchildren embarking on a day's outing. Sultan and Said were singing merrily, in between asking me questions and adjusting my headdress. Sultan wanted to know if I owned any horses, as he was keen on racing. Said enquired if I had ever driven a four-wheel-drive vehicle before, pointing out that the Sands could be very dangerous. They were so at ease and open that I knew we would remain friends for ever and my memory of the journey is of a great deal of singing and laughing. For me, it was the first time I knew that we would be able to reach the heart of the great Wahiba Sands in the Sultanate of Oman.

However, this story really begins a few months earlier on a very hot May morning of 1984 in the Conservation Adviser's office in the old part of Muscat with Roderic Dutton and Stephen Gilbert.

For those who know Oman, Roderic Dutton, Director of the Centre for Overseas Research and Development at Durham University, has become somewhat of a legend. He has been involved in agricultural and development projects there for some 12 years, with most of his energies directed to the setting up of a research farm on the Batinah coast. An Arabist, sensitive to development priorities yet concerned to maintain links with the past, he gave many in Oman the confidence to try new ideas that brought together the best of the past and the best for the future. This was appropriate for Oman, where change was rapidly taking place due to the increased oil wealth the country had benefited from since the early 1970s.

Dutton is not afraid to try out new ideas and technologies as long as they are of practical value, for his mission is to show how farming techniques can radically improve with new technologies, used in direct consultation with those who have traditional knowledge to share. Dutton believes that this knowledge, evolved by trial and error over centuries, must have a place in today's farming techniques, especially in developing countries.

Dutton had become a cornerstone to the planning of our project. He had visited Oman that February and by sharing the concept of the project with many of his key contacts, including the Conservation Adviser, his friends at several ministries and in particular the Ministry of Heritage and Culture, the Ministry of Agriculture and the Public Authority of Water Resources, had in a short time secured influential support. On the basis of this, he had offered recommendations as to how we should put forward our ideas on taking a multi-disciplinary expedition to the Wahiba Sands. All projects need individuals who create fertile ground for discussion and confidence for planning and Roderic achieved this as part of his overall approach to integrating development and conservation work. He is a true environmental ambassador whose visions will become increasingly important by the turn of the millennium.

His ability to turn problems into opportunities is a useful trait on an expedition. Blessed with a positive approach to anything that comes his way, he soon had the team saying, 'I think we have an opportunity here', every time something went wrong. A puncture would be used to create the opportunity to scout out a shop or market to order a chapati or two, to look at local produce, to fraternize and generally to learn more about the area. This philosophy was infectious.

Stephen Gilbert, the Chairman of the Project Committee, and I had been hit by the wall of Arabian heat as we descended the aircraft steps on to the tarmac of Seeb International airport. Even though it was not yet 8 o'clock in the morning, the hot air burned the inside of our mouths and nostrils and no amount of warning could have prepared us for the dryness. The airport buildings, however, were cool and welcoming as we were met by Roderic and Major Allan Malcolm from the Sultan of Oman's Armed Forces. The Army were very kindly offering accommodation during our brief visit and Roderic had arranged a series of key meetings, the first of which was with the Conservation Adviser, Ralph Daly, in just over an hour or so.

After unloading our luggage in the blistering heat of the garrison compound, we headed down the 40 kilometres of coastal dual carriageway that lead to the old part of Muscat. There is something about this apparently inhospitable and barren wilderness that exerts a fascination, as it has for generations of first-time visitors to Arabia Felix. The hazy dark brown mountains in the distance offer a stark and arid backdrop to the new empire being built along the coast. There is little sign of vegetation except on the dozen or so roundabouts along the fast dual carriageway, which are oases of dark green grass and beautiful flowers, fed by sprinklers. Buildings are sprouting everywhere and heavy plant and busy workers were much in evidence, striving to be ready before the National Day celebrations on 18 November.

The environmental and conservation world needs more Ralph Dalys and credit must go to Sultan Qaboos for appointing him Conservation Adviser at the Diwan of Royal Court. In the 14 years he has been in Oman, he has secured the respect and attention of decision-makers and local communities alike and in so doing has created a rare but important dialogue between politician and guardian.

Ralph is well known to those who come to Oman with environmental intentions. He has always been careful to ensure that any research is of practical value to the Sultanate and involves key ministries, and he has an acknowledged reputation for insisting that the rights of the host country are considered first. Although commonsense today, his edict was ahead of its time and his influence has been felt throughout the conservation world.

Throughout Ralph's career as an adviser he has acted as a catalyst who makes things work. Ralph has been responsible for the re-introduction of the Arabian oryx back to Oman after the last had been shot. The legendary success of the 'White Oryx Project' has given confidence and hope to many all over the world who have ambitions for re-introducing other animals that have become extinct.

Ralph received Roderic, Stephen and I in a cool, air-conditioned room and provided much-needed cold drinks. Armed with publications and management plans of previous projects, we put forward our proposals. Ralph confirmed there was a great deal of information we could collect that would be useful to Oman and began listing those with whom we should liaise in relation to the woodlands of the Sands, the night mists, the birds and animals, the plants and so on.

Long after our glasses were empty we agreed three things: first, that if we were to get permission to undertake this geographical expedition to the Wahiba Sands, we would incorporate the 'ecosystem approach', involving and integrating earth, life and human sciences, and that every effort should be made to ensure the results be of practical value to the long-term development of the area. Secondly, we must liaise with all the ministries concerned and in particular with the Ministry of Defence, who might be able to help with the logistics. Ralph said that if he could have our assurance that we would work closely with and for Oman he would look into the possibility of setting up an Omani co-ordinating committee so that representatives from all key ministries could be involved. Thirdly, we agreed that the project would not get any larger than 30 British and Omani scientists. The last was a promise the Society did not keep.

And so began the Oman Wahiba Sands Project 1985–1987. Ralph had made us feel very welcome. We had just 18 months to raise the funds, choose the team and put the show on the road. We went on our way with just ten days to follow up contacts and report to Ralph before we left. The sun was overhead and even hotter when we stepped out of the cool, white, thick-walled Conservation Office. Our shadows were very small indeed, yet our task seemed immense.

That evening we dined at the HQ Officers' Mess in the garrison known as Muaska Al Murtafaa (MAM). Covered in red and green lights, the national colours, this was one of four messes in the garrison and was to become one of many homes we established in Oman. Major Allan Malcolm introduced us to a number of officers either on secondment from the British Army to the Oman forces or operating as mercenaries on contract service, a number of whom had experience of the Wahiba Sands. The Sands are about the size of Wales but very few of the soldiers had travelled throughout them, most merely undertaking a single journey from north to south down one of the three main Bedu tracks. We learnt about the large dune ridges that dominate the north, the rolling white coastal dunes where wolves have been seen, and how during the summer months the winds are so strong that all vehicle tracks are blown away; how throughout the Sands Bedu encampments can be found, all of which offer hospitality, food, water and shelter; the fact that there are many wells but only a few marked on the maps; that at night the mists can be so heavy that everything is soaking wet in the morning. Also, of course, about the large areas of woodlands, particularly on the eastern edge, that are a refuge for animal and bird life. Gazelle sightings were regular and there was even a rumour of an animal resembling a donkey with zebra-like stripes on its hindquarters.

The Officers' Mess was to become a goldmine of information. Throughout my time in Oman I spent many hours here, discussing the project with soldiers who had collected a valuable body of knowledge during their time on exercises in the desert. There is a natural interest in the environment in Oman, particularly on the part of service personnel who have to spend much of their time on exercises. Oman's landscape is varied and vivid, sprinkled with pockets of strikingly beautiful oases. The Wahiba Sands were of growing interest and, due to better roads, had become more accessible. A recent press cutting in the *Oman Observer* featured a group of oil expatriates who claimed to have just made the first crossing of the Wahiba Sands. This caused a wry smile in the mess as it was known that a number had crossed it before on foot, by camel, in convoys and even with 10-ton trucks. However, few knew the Sands intimately and there were differences of opinion when discussing how dangerous they are.

Certainly they are hostile in the summer. Some of the sand storms can be fierce and those who have been caught in one describe a desert snowscene, where everything is covered in a fine layer of sand, just like an overnight fall of snow. And it is hot, often reaching 50°C. If you foolishly take hold of a steering wheel without gloves on, you will surely get burn blisters unless you have very hardened hands. Everything they say about frying eggs on the bonnet is true. However, we would only be there from November to April – the six coolest months. Getting lost was more of an inconvenience than a danger. As the Sands have finite boundaries – the mountains to the north, the sea to the east and dried river beds to the west and south – it was difficult to become properly lost as one might in the Rub' Al Khali – the Empty Quarter – or the Sahara. This was one of the values of this particular sand sea and the reason it was safe to let researchers enter it relatively unleashed.

There was one officer whose reputation had gone before him whom we were pleased to encounter. Raa'id Chris Beal or 'Bealsky' as he was sometimes known due to his Polish–Bavarian ancestry was not only knowledgeable about the Sands, having crossed them several times, but was an experienced expeditioner, with some hair-raising trips in the forests of Malaysia and a ballooning expedition across the Sudan under his belt. The thing he understands best, however, is transport logistics. He was second in command of the Transport Regiment and I had been hoping I might be able to get him involved in the project as an adviser.

Bealsky looked tanned and fit, having just come off exercise. He fortunately took an instant liking to the project. When we explained we wanted to get an overall picture of the area which must include those remoter areas where the sands are soft and few Bedu live, he immediately drew an outline of the Sands on the back of an envelope. He told us that Mintirib would be the best location for setting up a central base as there were routes directly south between the fingers of the dunes, to the west following the Wadis Andam and Matam and to the east following the Wadi Batha. It was Chris who, both now and at later discussions, gave me a true idea of the scale of the operation and why it would be useful to have substantial back-up if things did go wrong, including constant radio communications, driver training for the team, sub-camps with fuel and water depots, special route-marking pickets and so on. He pointed out that although the Sands are relatively small and the Bedu usually around to help, getting vehicles stuck in the larger sand dunes would waste valuable time and, if the vehicle rolled, could even cause a fatal accident. Slowly some priorities began to emerge. A scientific reconnaissance to choose the main areas of work and a period marking the main routes would be essential. Flights over the whole Sands would be exceptionally helpful.

It was now late. As Roderic, Stephen and I walked across the garrison to our air-conditioned cabins we reflected that this advice and encouragement from the services was important. In Oman the Armed Forces dominate quietly and effectively; we would not be able to work in the Sands without their assistance.

I seldom lie awake but despite being four hours ahead of UK time I could not sleep that night. After a bout of tossing and turning, I settled down just after midnight at a small table under the light of the desk lamp and began to sketch the outline operational plans of the project on several large pieces of graph paper. Despite the enormity of the task when laid out before me on the paper it suddenly seemed so very possible and real. We had made our first moves. Not long before dawn I turned off the air-conditioner and climbed between the sheets. Some lines of Goethe came to mind as I drifted off to sleep: *'Whatever you can do or dream you can, begin it, Boldness has genius, and magic in it.'*

If we believe that expeditionary science breaks down barriers between cultures and provides fertile ground for discussion on how to tackle environmental and development issues, an assessment of the communities we are trying to influence must be made. By the end of our project, the Royal Geographical Society had made contact with over 1,000 individuals who were to play a role in it in some way. It was like a large orchestra with many musicians, all contributing a vital part. Within Muscat there were four distinctive communities to consider, all of whom we needed and wanted to involve. Involvement means discussion. For discussion it is important to have a '*qahwa* and dates' with individuals receptive to the ideas. To make an appointment for *qahwa* and dates requires an introduction.

The first group were the government ministries. Liaison with representatives was crucial. If we failed to involve the decision-makers responsible for the future welfare and management of the Wahiba Sands then we would have failed as a project. It was as simple as that. With steering advice from Ralph Daly, a special co-ordinating group was to be formed comprising representatives from the 20 or so different ministries. Supporting us here was our own British Embassy and the British Council.

The Armed Forces were the second of the communities. From the moment Allan Malcolm introduced us to the Army headquarters and to characters in the mess it was clear our links would grow. Letters of support had also gone from our own Council to senior representatives in the Ministry of Defence.

The third group centred around the thriving business community and we were to liaise closely with several of the major companies in Oman keen to help environmental research. The immediate assistance of eight companies who became the main backers of the project, called the Corporate Patrons, enabled the project to plan ahead with confidence and determination. The wider visions offered by many companies such as British Petroleum Arabian Agencies helped us look at our educational responsibilities and we became involved, in conjunction with the Ministry of Education, in producing literature in Arabic about the Wahiba Sands for Oman's schools.

The final group was the scientific community. We were most fortunate here. The magnificent new Sultan Qaboos University would be opening almost at the same time as the project began and it would provide opportunities for people to meet and discuss matters relating to the Sands. This would ensure we had liaised with all interested parties within the capital and it would be the university that would host the final presentation of the results.

So our strategy in Muscat was based upon bringing together these various groups to share knowledge and expertise. Although some of the seeds sown might not necessarily bear fruit for another ten years or so, a partnership between science, commerce, services and ministry representatives is a valuable way to do justice to an area like the Wahiba Sands. By focusing on it as a geographical feature, the future priorities of a region can be established.

Expeditions are about meeting people – committed individuals who are prepared to offer their talents unselfishly as part of the whole. We were soon to learn that all four communities contained these – in fact, within Oman there is a vision and hope that goes far beyond tomorrow and we found a very positive response to the concepts of collecting pieces of the geographical jigsaw of the Wahiba Sands for the future. There are good reasons for this.

The first is linked to history. Since the early seventies the Sultanate has benefited from a large investment of oil revenue. Under the leadership of the Sultan Qaboos bin Said, Oman has made some very long-term plans with care and caution, the kind of development denied to many countries. By learning from the mistakes of others Sultan Qaboos has taken every step possible to integrate conservation and development, and his visions are ten years hence rather than anchored to the present. Care has been taken at all stages to think through the implications of any development programme, and his respect for the environment has been noted by conservation bodies throughout the world; he could be described as one of the world's greenest statesmen.

Behind this vision is a traditional and religious respect for the environment. The Islamic faith has a strong and clear policy towards the 'conception of the universe, nature, natural resources and the relation between man and nature'. Indeed, the Koran has many references to respect for and protection of the natural world. To quote Sultan Qaboos:

> Development is a necessary process, especially in our country, but the impact of development must be skilfully directed so our woodlands are not destroyed, our soils are not eroded and our valuable species of plants and animals are not prevented from playing their vital role in maintaining the environment, given us by God, in which to live. Our plans for development must be based on facts; facts about resources, our environment, our ecosystems and facts about how we, as human beings, exist in interrelationship with the wild plants and creatures who share God's earth with us.

As far as we were concerned the Sultan's far-sighted policies provided fertile ground for geographical research and so helped us bring together the four main groups of Muscat.

One of the first meetings Roderic had arranged was with the British Embassy, which is situated in the centre of the old walled city of Muscat. A visit to the old city provides a rich whiff of history that begins as soon as the two imposing forts, set into the mountains on either side of the bay, come into view. Forts Mirani and Jalali, built by the Portuguese in the fifteenth century, look menacing as they sit watching events unfold below. They have seen bloodshed. The Portuguese, occupiers of Muscat for 150 years, considered them impregnable to the extent that in 1650 the then Portuguese Governor offered such provocation to the ruler of surrounding Oman as to send him a piece of pork in response to a request to purchase provisions in the market. This sparked a fierce attack by the Omanis which overthrew the Portuguese, although a small number held out in Fort Mirani for several months. The price was heavy; it is believed some 4,000 Omanis died in this battle. Until recently the forts were used as a gaol, but they have now been restored and make a noble feature on the otherwise naked skyline.

The Embassy is on the seafront next to the Sultan's Palace and from its courtyard overlooking the bay has a commanding view of the forts. It is an impressive white building, protected by two huge Arabian doors, and was built in the later part of the nineteenth century. Inside, the overall aspect cannot have changed much and the past is recorded by a set of photographs of former ambassadors lining the stairs. I noted Sir Percy Cox, ambassador in 1890, whose maps and photographs I had seen when delving into the RGS archives back in London. Links between Britain and Oman remain as strong as ever, illustrated today by the number of British personnel in the Oman Government who have had a major influence on the design and structure of the services, the police and the university.

The Embassy, although small, is exceptionally active and very hard-working, with a high turnover of businessmen, advisers, VIPs, royalty including the Prince and Princess of Wales and, increasingly, scientists involved in research and teaching within the Sultanate. Richard Dalton, the first secretary, himself an Arabist who had already succumbed to the spell of Muscat and extended his tour, was keen to foster further scientific and educational links and to help make the right introductions for us. It was important that Richard understood our intentions as no expedition can ever be undertaken without the closest co-operation from diplomatic representatives. Expeditions provide fertile ground for misunderstandings and diplomatic steering is crucial.

The days were hectic. I was glad to have the support of Stephen and the introductions from Roderic. Our list of people to meet included Kamal Sultan, President of the Historical Association and also a senior director of W.J. Towell and Taylor Woodrow-Towell, two leading Omani companies. Kamal Sultan is a legendary figure who has a very special interest in the natural resources of Oman. A tall Omani with fine, distinctive features and a friendly and inquisitive face, he is passionately concerned about the environmental future of the country and, as a senior politician on the Consultative Council, his views were to be respected. Even before I had finished telling him about our plans he had started to give me advice and help. His company would certainly lend support, he said. We must talk to John Cox of Taylor Woodrow-Towell. And, of course, he would have a party for the team. Hospitality is the middle name for the Sultan family and he entertained the whole party on several occasions. I came away from Kamal's office absolutely reeling with ideas; I later learnt that Kamal is Oman's answer to Kurt Hahn, the founder of Outward Bound, and is a great believer in outdoor and environmental education.

Perhaps one of the most significant meetings was with Carroll Hess, a cheerful American from Kansas. Carroll, who was to be a prop-forward for the project, was responsible for forming the Agricultural Faculty for the new university and in that capacity was an acting dean. In his turn, he introduced us to Sheikh Amor bin Ameir.

Sheikh Amor is a wise and experienced gentleman. To be the Vice-Chancellor of the university, he now headed the team of deans responsible for establishing it. It was no mean feat. When we met in his office he was surrounded by charts, reports and technical specifications and how he could calmly stop for ten minutes to hear out our plans I do not know. There was a peacefulness and depth to Sheikh Amor that many of us have since commented on. Formerly Minister of Education in Zanzibar, Sheikh Amor had always been involved in education matters and here was the most important task of his life – the formation of a brand new university. Roderic, Stephen and I confirmed that we wanted to work closely with the university, that we hoped there might be some scientific collaboration and that we might present the results of our work at a special symposium in 1987. Sheikh Amor smiled and, although without commitment, offered his support for the project and explained that Carroll would be our point of contact. He then wished us all well for the aerial reconnaissance the next day.

Stephen's request to the Ministry of Defence for an aerial survey had been received with enthusiasm and a SAF Defender eight-seater was ready to fly us on a reconnaissance around the Sands. At 6.00 a.m. the next day Carroll, Stephen, Roderic and I stood on the tarmac of Seeb airport armed with satellite images, maps, cameras and water bottles. This was to be our introduction to the Sands. Our pilot, Rashid, was a young Omani whose home town was Mintirib, the town at the northern tip of the Sands, so consequently he knew the area well and could recognize the areas of woodland that stood out on the satellite image. We asked if we could follow the dry river bed called the Wadi Batha that coursed round the northern and eastern edges of the Sands and down past the woodlands.

Circling over the deep blue Indian Ocean, Rashid headed south east over the stark yet striking Hajar mountains that lie between Muscat and the Sands. The Sur road snaked its way through the passes, occasional clumps of date palms indicating small settlements and villages. The ground looked rough and barren. The courses of the young rivers could be seen carving a route down the mountains and sudden flashes of sun indicated a few small pools where the rivers had been dammed. Within 20 minutes, however, the mountains gave way to the plains of the dried river beds, which widened out into distinctive fan patterns disappearing into the distance. With no rain now for over three years, fast trucks following the gravel roads created large dust clouds that hung in the air.

This area is known as the Sharqiya, or the 'eastern region', and extends to Sur on the coast, encompassing the Wahiba Sands. One of the major provinces of Oman, the Sharqiya has historical links with East Africa and in particular Zanzibar, where the Arabian influence is said to have been left by the Al Harthy tribe, who dominate the Sharqiya. Many Omanis fled Zanzibar during the revolution in 1963 and headed back to their homeland to settle in the villages of the Sharqiya, including Qabil, the Al Harthy capital. These villages could be seen as splashes of green blending into the dark brown pattern of houses. Tell-tale rings of the *falaj* drill holes formed a line of craters in the wadi, evidence that a water channel was taking fresh water from the wadi underground to one of the villages. The absence of rain had left most of the vegetation, sparse as it was, looking very parched.

In the cockpit Rashid was peering ahead, indicating he could now see our objective. In the distance the Wahiba Sands appeared as a thin pink line cutting across the horizon. Before us lay one of the few isolated sand sea deserts of the world – a great expanse of mobile sand kept in place by the rivers and the sea. At first it was difficult to believe it extended south over 240 kilometres, but even at this distance the ridges and valleys could be seen, resembling the fingers of a giant hand laid flat on the ground. Dunes of darker red sand were piled against the mountains to the north, a reminder that the sands are continually on the move. Over many thousands of years some part of them has managed to cross the gauntlet of the Wadi Batha which effectively keeps the main body of the Sands in check.

Where the grey wadi cuts into the dunes, a wall of pink sand with steep slipfaces indicates the start of the sand sea. Rashid had brought the plane down to just 150 metres so we could now see it very clearly. Each of the fingers was a high ridge of dunes rising nearly 91 metres above the plain, pointing in a southerly direction. Between each was a swale some 2 kilometres across, shallow valleys extending deep into the interior of the Sands, lost today in a pale haze. Each seemed to be covered with a stubble of vegetation that looked dried and withered, with a distinct track running down the centre and various Bedu camps made from palm thatch dotted around.

The characteristic 'Wahiba pink' of the sands is worth comment, for there is in reality an infinite diversity of colour creating an illusion of continual change. The sand, comprising a mixture of broken rock particles from the mountains and sand from the sea floor, varies in hue from snow white to jet black, although the predominant shades range from pale honey to the colour of dried blood. In the morning and evening sun the dunes can be as bright as flamingo pink, but during the day they become pale and hazy. An artist commissioned to capture this pink complained that the sands changed in colour every ten minutes or so like a slow-moving slide show. Although this effect is beautiful and atmospheric, more importantly it indicates the great variety of minerals to be found in the Sands.

Rashid had limited fuel for this reconnaissance so we carried straight on towards Mintirib, passing over a number of villages, collectively known as Badiyah, that are found at the north end of the Sands. As we flew over Mintirib we asked Rashid for a closer look at what might well be our main base. He immediately went into a nosedive that nearly took us into the marketplace, missing the tops of the tallest date palms by inches. Below we could make out the old mud fort in the centre, flat-roofed square houses with walled gardens, the high street with a variety of modern shops, a bustle of people and a number of trucks in the market, and a dense cover of date palms. Camels, sheep and goats cast a wary eye as we skimmed overhead. A group of riders on horseback were having a race down an improvised track in the wadi. That was Rashid's thumbnail sketch of a desert village.

We continued to follow the wall of sands eastwards and then in a great arc across the centre. The flight confirmed the diversity of sand and dune types. The north/south ridges typical of the Wahiba were distinct and, superimposed on the top of these long ridges, I recognized crests and waves I had last encountered in the Sahara, and barchans (crescent-shaped dunes) and patterns I had seen in books. How could such a medley of different dune types be found in such a compact area – and why was the sand such a mixture of different rock materials?

Throughout the Sands vegetation was omnipresent, albeit quite sparsely in places. Even in the recently formed and soft dunes by the coast, plants were established and at first glance the variety seemed greater than the 18 or so species that had been recorded there to date. The most striking feature was a dense band of woodlands which we followed down the east side; they ran in a long thin line for over 80 kilometres. This desert forest comprises a remarkable tree called *Prosopis cineraria,* which is indigenous to Oman. The prosopis woodlands, on the eastern and western margins, stood out clearly even from a distance while a greening of vegetation covered the ridges and swales alike. We saw goats, sheep, camels, and a few gazelle and *barusti* settlements, often with a bright-red Toyota pick-up parked outside, which confirmed that quite a substantial population of Bedu lived in the Sands. Scars of earlier explorations crossed the area regularly from east to west, remnants of the rentises (tracts) cut by seismic crews for oil exploration now being covered in windblown sand. All the other established tracks followed each of the swales south for some 50 kilometres or so.

The woodlands gave way to the white mobile dunes by the coast with a few fishing settlements right on the shoreline. The Arabian Sea was dark, dark blue and small fishing boats bobbed on the waves while large flocks of sea birds living off the debris of the fishing communities glided below us. At the southern end there was a large *barusti* settlement – a fishing station with a fleet of boats on the shore, nets drying in the sun and an assortment of trucks with large tanks for carrying fish on the back. To the south was the Bar Al Hikman peninsula, where there was a profusion of migrant wading birds including flamingoes, and to the east was Masirah Island. We turned westwards and then northwards and followed one of the central swales right up through the Sands. As we neared the northern edge, Rashid flew low down the swale so that the crest of the dune was level with us. The colours, shapes and patterns defied description. We passed the sharply defined edge where the wadi trims the sand and suddenly we were back over the Wadi Batha and the Sharqiya plain.

The flight lasted two hours and a confusion of images were left in our minds. Where had all the sand come from? Why was there such a variety of patterns and colours? How did so much vegetation endure the lack of rain? How did the Bedu survive and why did they stay in the Sands while the rest of the Sharqiya moved into the twentieth century?

For Roderic, Stephen and I these questions confirmed there were many frontiers here for a geographical team. The Wahiba Sands are considered unique by the desert experts, since they have the greatest diversity of sand types and formations anywhere in the world. However, with so much vegetation and with evidence of a large Bedu population making them their home, it is also the site of a living and thriving community with an inbuilt desire to resist change and conserve the ecological status of the desert. Assessing the dynamics of how this ecosystem is maintained would contribute greatly to the scant but growing pool of knowledge about deserts and might offer valuable lessons for other parts of the world.

CHAPTER TWO

Preparing for Arabian Sands

The UK press are as fickle as the weather. Getting them to take an interest in environmental issues overseas is a hit and miss affair; much depends on what else is on the agenda that day. On Tuesday 3 September 1985 we launched the project with a major press conference and reception.

At 11.00 a.m., the car carrying His Royal Highness Prince Michael swept into the gravel drive of the Royal Geographical Society. The line-up shuffled to get into place and peered towards the front entrance as the Prince strode through the heavy swing doors, now safely held back on their latches. This was it. Our Patron had arrived to inaugurate the project officially and there was no turning back. Sir Vivian Fuchs, a past President, and John Hemming, Director of the RGS, welcomed the Prince. Cameras flashed and I wondered if these pictures would make the newspapers the next day. Fleet Street had never been so well represented at an expedition launch; some 30 or so journalists had managed to fit this into their diary but it was, after all, the first time an RGS expedition had had a royal patron at the helm. His Excellency Hussain bin Mohammed bin Ali, the Omani Ambassador, was introduced at the head of the queue. The beaming face of our Patron was most reassuring and my launch-day nerves quickly disappeared. Cameras flashed yet again as I took the Prince down the line of key project members, sponsors and guests from Oman. Two Gulf Air hostesses were waiting at the end of the line with some Arabian dates and coffee for us as we collected together in the tea room; Gulf Air, who had kindly offered to help the project as the official carrier, were doing more than just providing seats on planes.

Meanwhile, round the corner over 200 guests and press had arrived to join in the launch. They too were being offered refreshments and were having a look at some of the displays of equipment, photographs and maps that had been prepared and at a unique display of Omani jewellery, *khanjars* and silver that had been specially flown in for the occasion. Smells of roasting goat wafted from the garden through the house. There was, for a short time, quite a festive air. I came down to earth as someone pulled at my sleeve to indicate we could now go and face the press. This is always a tricky moment.

I had to get press coverage today; failure to do so would severely hamper our fund-raising effort. Furthermore, it might mean that the eight main sponsors, or 'Corporate Patrons', of the project would not be seen to be identified with it and questions would be asked at their respective boards. Funds might even be taken away. If our Omani contacts were to believe that we considered this an important project, it was vital that today's events reached the media both here and in Oman.

Many people question the value of speaking to the press on the basis that trying to secure news coverage is very time-consuming, rarely produces results and creates wrong impressions. 'Let the final results of the project speak for themselves' was the opinion of the senior members of the team. However, the results of any research project should reach a wider audience than those who are professionally involved. Scientists are continually castigated for not being more effective at communicating their findings to those who need to understand the issues and whose actions will be influenced accordingly. To my relief every national paper was represented and there were many other reporters with a particular interest in the Middle East.

We filed into our main lecture hall, which was nearly full. The Omani and Union flags were draped on the balcony and a huge Gulf Air banner with the title 'Oman Wahiba Sands Project' in Arabic and English was strung between the two. Sir Vivian Fuchs, Roderic Dutton and I took our seats on the rostrum.

After our presentations the floor was opened to questions and of the many that were asked there was one to which everyone wanted an answer: 'Will the project provide solutions to the current drought problems in Ethiopia and Sudan?' Roderic took the helm, as this was a key issue. He referred to the United Nations Conference on Desertification in Nairobi in 1977 where a number of master plans were discussed. The urgency for a global strategy was unanimously agreed and 28 specific recommendations made. There was hope that by better co-operation between nations this initiative would halt the tide of desertification, but today's estimates suggest that some 60,000 square kilometres of productive land are lost to desert each year and there is little hope for the future. No, we would not be coming up with answers overnight that others have failed to offer, but we would be adding to the growing pool of knowledge about arid and semi-arid zones that all future managers and decision-makers will have to use to assess priorities. As ten times the population of the British Isles live in these regions, it seemed to us that there is a degree of urgency here.

It was our belief that by fielding a multi-disciplinary team of geographers we would produce a balanced statement which would put the Wahiba Sands firmly on the scientific map. At present, desert research is considered to be low priority and, globally, little money is spent on it for a variety of political and economic reasons.

Roderic went on to explain why this project could lead the way for future investigations and so indirectly assist Ethiopia and Sudan. Global strategies are fine so long as there can be dialogue on the ground with those whose lives are being affected. In practice this is not easy. Furthermore, different regions of the world have different ecological components and international policies may not be able to cope with the varying types of environment. Ecosystem studies, bringing together scientists of several disciplines who are able to talk to both the local guardians and the key decision-makers in government, can give a useful understanding of the processes of the area and thereby predict and plan for the future. This was our hypothesis and we knew in the long term it would contribute to the task of turning the deserts green.

Preparing for the launch had not been a picnic. In all projects it is important to incorporate cornerstone dates by which certain things should have happened. We were due to go into the field by early December, so with only three months to go it was important we were on target. Apart from the day we were to fly, this was the most crucial of all our dates. We had used it to focus our planning priorities, which primarily included at this stage choosing the team, raising the finances and securing the field equipment. And, for effective planning, you need an effective administrative team – Wahibadmin.

There is never a shortage of volunteers who are prepared to contribute to scientific expeditions. The support team comprised a diverse group, some of whom I had known for many years and some who had literally walked in off the street.

Mike Holman, an ex-Royal Marine who had worked in Oman before, arrived in my office one day. Fit, keen and competent, he immediately fitted the bill and he took on the responsibility of the vehicles and radio communications – two cornerstones to our operations in the field. Rebecca Ridley likewise came in off the street looking for a job and ended up as my assistant. An Exeter geographer who had served in Gambia with VSO, 'Beck' was particularly interested in the educational aspects of the project. Ever cheerful she became the administrative manager for the team, keeping the growing paperwork under control as well as looking after the needs of the team as a whole. Both Mike and Beck were keen to work in Arabia and it is not surprising that both were to get jobs in Oman after the project.

I needed a medical officer and rang Nicola Bennett-Jones. The best expedition nurse in the world, she is tough and beautiful. We had worked on two earlier projects and I knew her worth. She had mopped my fevered brow in rain forests and dressed my wounds by an African river. She could cope with emergencies large and small as well as run a base camp on her own. To my delight, she immediately agreed and set-to sorting stores and supplies.

I needed a base camp manager. I rang Mike Keating, a Cambridge historian with whom I had worked on the Kora project and committed to environmental development issues. Currently in City publishing, Keating couldn't join for the whole time; he had been offered a post as Sadruddin Aga Khan's personal assistant in Geneva. I couldn't compete. The Prince had agreed to pay him a salary no less! Nevertheless I did secure Mike as a photographer – one of his many professional talents.

I was now lacking a field director. In Nairobi in 1983 I had met Nick Theakston who was working for the management consultants Peat Marwick McLintock and who had been most helpful to our project there. His interest in the Society and his expertise at keeping a firm eye on the overall management and accounting of the project was a talent I needed. I had no idea if Peats would second Nick to the team. Certainly the expedition world is striving towards more professional administration; I saw the chance to benefit from a partnership with a progressive company which would not just loan Nick but would design the management structure of the project as well as keep a firm eye on the finances. Peats displayed their true environmental colours and agreed to Nick joining the Project for six months. So the Wahibadmin team was formed and took on the responsibility of the overall safety and welfare of the project. Above all the project had a cheerful multi-talented team of unselfish environmental crusaders who were all prepared to work hard to make it succeed. The RGS was lucky. Today they were particularly busy.

For my part this launch was also important for our partnership with our financial sponsors. Fund-raising for an expedition is always a headache. Traditionally funds come from a variety of sources, but with current cutbacks to government funds, bodies such as the National Environment Research Council have little to spare. Other grant-giving bodies are also short of cash so corporate support from Britain and Oman was vital. We needed a minimum of £210,000 to make the project work. With over 4,000 accumulated days in the field, this amounted to less than an overall cost of £50/person/day including full board at Taylorbase, travels costs to Oman, field equipment, equipped Land Rovers and all the associated administration. We considered this to be an investment and told commerce so.

In return we wanted to show a partnership between science and commerce. To become a senior sponsor or 'Corporate Patron' as we referred to them, we asked for a contribution over £20,000. For this, prospective candidates received front-line association with the project, an opportunity to visit the team in the field during the visit of our Patron and a variety of other promises relating to PR and mentions in the media. But we stressed it was to be a partnership. As it turned out our final sponsors were very unselfish, giving far more than they were to receive back. A full list can be found in the acknowledgements.

And so I began my exploration of the business community in Muscat, which was to be the turning point of the project. My most significant meeting was with John Cox, the chief of Taylor Woodrow-Towell. 'Here is another old hand who will send me packing,' I thought, but my hopes were raised when I noticed a signed print of Ran Feinnes and Charlie Burton at the North Pole. I launched into my piece about the project and why our work would benefit Oman, humankind and world knowledge. John looked intently at my portfolio describing the area, our timetable, our operating plan and the budget. He didn't beat about the bush. 'How can we help?' he asked.

'I wonder if you might be able to construct a research base for us somewhere near Mintirib, at the north end of the Sands?' I asked. 'We will need a headquarters to serve the project for four to five months and to accommodate up to 40 people.' I was about to launch into the merits of being a Corporate Patron and what we would do in return, but I didn't have time to finish.

'I am sorry, I have to go now, but I am pleased to say we will help you build your field university – let us know what you would like in due course,' and on that note he bid farewell. My head was reeling. The whole meeting had been less than ten minutes, but I was to learn that that is how John works and we never looked back. Taylorbase had been born. This initial promise of support from John Cox was to shape the whole structure of the project, as I now had a base camp worth many thousands of pounds. This promise so early on would certainly help our contact with other sponsors. I quickly marked in TAYLORBASE at Mintirib on my map and looked to see who was next on my list.

The members of the business community whom I visited all offered tremendous hospitality and, after dates and coffee, kindly listened to my story. There is a tremendous pride in Oman's heritage. The diversity of geographical features within the Sultanate is remarkable and a genuine interest in the Wahiba Sands and its night mists helped my case. I focused on the multi-disciplinary angle, on our educational responsibilities, on the better use of technologies, on the knowledge of the Bedu nomads and, of course, on a huge magnificent satellite image of the Wahiba Sands. Having seen that image unfold on a table, it is not easy to ignore the Sands.

By the launch of the project eight companies had pledged support in cash or in kind in excess of £20,000 each and five of these were Omani businesses – namely Assarain Enterprise, Suhail and Saud Bahwan, Mohsin Haider Darwish, Taylor Woodrow-Towell and Zubair Enterprise. Land Rover (UK) was prepared to loan five brand new 110 Land Rovers and Racal Tacticom to equip us with eight amateur-proof, variable frequency radios. To all, I had said that we would be launching the project on 3 September and that we should get some international coverage in the press.

So now, as we munched roast goat kebabs in the RGS garden, I spared a thought for those who had so gallantly committed themselves to the project and who had given us the confidence to be ambitious with our planning. A selection of the equipment ready for use in the Sands was on display – heavy-duty sand-ladders to get us out of soft going, special protective camera bags, field computers, lightweight video cameras, a portable printing machine with special ink for the tropics, a hand-held data-logger and mats to stop us sleeping on scorpions.

British commerce has a long history of supporting expeditions in kind, but the sponsorship deals are often a little one-sided and projects do not give back to those companies the recognition they deserve. A quick browse through our acknowledgements list will indicate the kind of generous and unselfish support given to scientific endeavour. We were proud to be able to show Prince Michael and all those who came to the launch some idea of this magnificent commitment.

Team members lined up to say farewell to our Patron and his private secretary: 'See you in the Sands.' Everyone was very relaxed and we were looking forward to showing Prince Michael our fieldwork. We could now get on with the final detailed planning in Oman and within a few weeks we would have Taylorbase complete and the project operational. Would any of the 30 or so members of the press who had come be able to find space for us within the papers the next day? Enough film had been exposed to stretch from London to Oman. We were hopeful and felt the launch had gone exceptionally well. The morning had started with three radio interviews – for LBC and for the *Today* programme and *News at One* on Radio 4.

On Wednesday 4 September 1985, not one national newspaper carried the briefest mention of the launch.

CHAPTER THREE

Taylorbase – the Field University

For all expeditioners Wilfred Thesiger is something of a hero. He is the last of the great desert explorers and for many he crowned the work of Burton, Doughty, Blunt, Lawrence and Philby, familiar names for those who have taken an interest in the British presence in the Middle East. The British have always had a soft spot for deserts and the RGS *Geographical Journal* is full of detailed reports by many less well-known explorers who toiled for months and sometimes years to bring back geographical data. A common factor is their respect and appreciation of arid areas and of the opportunity to share them with those who have made such places their home. Arabian culture has its roots in the desert. However much you may have read, however, there is a whole dimension to being at peace within a desert that cannot be adequately put into words. Wilfred has come close and, through his work, he has inspired many to follow him. But, as he quite rightly pointed out, it is no use just dreaming – his advice is to 'go and have a look for yourselves'.

I met Wilfred Thesiger in Sarawak back in 1977 when he came to visit us at the self-built longhouse which served as our research base and, in the evening, held us spellbound with tales of his travels. It was the first time I heard about the great Wahiba Sands and I can remember quite clearly his descriptions of the strangely coloured dunes, the camels with which he had a love-hate relationship, the hospitality of the Bedu and his profound affection for them, and his fear that their knowledge of the desert was being lost. 'They are masters of the desert environment and they live there because they choose to, not because they have to,' he said. Looking back to his time with his two Bedu travelling companions, he remarked that although he had seen some of the most magnificent scenery in the world and lived among tribes who were interesting and little known, none of those places had moved him as did the deserts of Arabia.

Some nine years later in 1986 I was now standing within sight of the same dunes that Wilfred had camped in. They overlooked the Badiyah area and the villages of Mintirib, Raqah, Sharhiq and Wasil. This site, on the track leading to Hawiyah just 2 kilometres from the centre of Mintirib, was to become our headquarters for the next five months. It was difficult to believe. The corners of our camp were marked out in the sand and the holes for the foundations were to be excavated today. I sank into the sand and dug my hands in while I watched the Taylor Woodrow-Towell team start work. This was the beginning of Taylorbase.

Tony Hyde, the site director, would have been an excellent expeditioner. He knows all there is to know about construction work in remote locations and how to get the job done. Action and not words is his motto. I met him in the tent that served as his office, where there were charts and files strewn around, a radio making noises-off and a general air of progress. The paraphernalia of construction lay round about and the bustle of some 70 camp builders gave the impression of a film set.

With me was Captain Chris Griffiths, a member of the Sultan of Oman's Armed Forces who was to head up the operations side of the project. He was the training officer for the Coast Security Force, whose area of patrol included the Sands, and he and his soldiers probably knew the area better than any of the other services personnel in Oman. Based at Ibra, he had spent many hours driving off track through dunefields. Known as 'Kriss' to his soldiers, he was a popular figure and had a reputation as being a bit of a daredevil. He had generously taken time off to escort me to the eastern part of the Sands to reconnoitre and agree a second base from which we could operate.

Tony Hyde took us round the site and described what would go where. Although he did not say so, he clearly thought we were bonkers but he didn't let us down and, with a large team of Pakistani and Indian constructors, completed the camp in just two weeks. The force worked day and night under huge arc lamps while Tony seemed to keep at it without a break.

Kriss and I were to return in two days but for the moment had an urgent decision to make concerning Field Base, the second operating site. All others would be mobile bases and would move on every two weeks. Field Base, or Qarhat Mu'ammar as it was to be locally known, was to be near the eastern margin of the prosopis woodlands and also close to an area of mobile dunes that were suitable for study. We followed the Wadi Batha round the northern and eastern edge of the Sands towards a small hill with jet black stones on the top.

Driving one of the army V8 Land Rovers I followed Kriss along the wadi, which follows a course east and then south. The high mega-dunes looked forbidding; it was this great wall of sand that had kept so many people out of the desert. Through his training operations with the Coast Security Force, whose area covered the majority of the Sands, Kriss had accomplished during the summer a west–east crossing, travelling perpendicular to the ridges, thereby proving the journey feasible. Many pundits had said that it would be impossible without large logistical support, air drops, etc., but Kriss's confidence in knowing how to travel through, round and over high dunes, both soft and hard, was to become an important element in assisting our team to follow in his footsteps and in turn gain their own confidence.

'We shall just nip up here,' he said, with a gleam in his eye. I wasn't keen to hesitate at this stage. We let our tyres down to 15 PSI, I by using a tyre gauge, Kriss by giving the tyres a kick with his boot. He took off like a jack-rabbit straight up the fierce incline through what appeared to be extremely soft sand.

I followed in Kriss's tracks, giving the engine plenty of acceleration as we raced up this slipface, round a corner and up two further inclines. My confidence grew as the soft radial tyres padded over the sand and I began to shout with excitement as the vehicle went higher and higher up the dunes until I was nearly 75 metres above the wadi. One final ridge and I would be at the top. I could see Kriss waiting. Two seconds later I was stuck up to my axles, with the tyres digging in every second. I took my foot off the accelerator and the engine kicked back violently as it settled down. I realized how hot the engine had become as the smell of burning oil and grease wafted up through the dashboard. Rule number one is to reverse out of the problem, but I had a dilemma – I was on a ridge. I was tilting sharply to one side and it was touch and go whether I could steer a safe course; the drop to either side was alarming.

I took out a spade and dug under the wheel at the top of the slope to balance the centre of gravity slightly then climbed back in and very gingerly reversed to a flatter ridge about 5 metres wide. I should have then backed the Land Rover up the slope opposite to give me a run at the ridge – an easy trick we were all to learn. But I did it the hard way; I made it on the sixth attempt and arrived at the top with sweat dripping down the sides of my face.

On the other side the dune sloped away sharply. It was the slipface – the side the windblown sand falls down, forming a wall of sand usually at about a 33° angle. I began to have visions of the team flipping their vehicles on these slipfaces on a regular basis, but Kriss assured me it wouldn't be difficult to get the whole team used to such dune driving.

He took off again, this time going straight down the drop. It is always worse than you think because the bonnet sticks out such a long way and the tyre on it blocks the view. 'Low gear, keep straight and no brakes!' he shouted as a parting reminder. Over I went. It was an alarming angle but there was no going back. Halfway down I hit a clump of bushes. The vehicle slewed to the left, for a second I touched the brakes and the back began to slip sideways even further. Instinct made me accelerate out of this as one does out of a skid and I reached the bottom of the slipface with very white knuckles.

After this brief introduction to the dunes I began to think about getting the team ready for the Sands and the three days Mike Holman had planned for driver training. It was going to be very important that all members of the team became used to driving in the Sands both day and night, with heavily laden vehicles. From top to bottom of that dune it was some 45 metres and if Kriss had said beforehand I was to come down that, I would have told him what to do with the safety flares. Like parachuting, though, having done it once you want to go again.

Now was not the time, however. We had a long way to go before nightfall and so we continued our journey along the gravel-strewn wadi. It was very dry and I drove in the dust cloud behind Kriss's vehicle as he followed the course of the wadi for some 100 kilometres. Much of it was grey in colour from sediment that had been washed down from the mountains. There was no water to be seen except in one or two areas where the wadi narrowed and bulrushes grew. This was suddenly quite a different world with dragonflies, birds and frogs, where we pushed through rich vegetation and splashed across large, shallow pools. Small fish darted under the Land Rovers as we went slowly through. Such places were rare as the river remained underground for most of the time. Although there hadn't been significant rains for at least three years, the wadis still had plenty of good, clean water that kept the many wells we saw topped up, and indeed Wadi Batha was well known for its sweet water.

Wadi Batha demarcates the Wahiba Sands and following it was an interesting way to reach the villages of Kamil, Bilad Bani Bu Hasan and Bilad Bani Bu Ali that are found down the north-eastern flank. We left the wadi just south of Bilad Bani Bu Ali, climbed up on to a flat plateau of sand and gravel and headed due south. Within a few moments we came across the camel-racing track which we were to use as a navigation aid. The prosopis forests slowly appeared to our west as we sped 60 kilometres towards the 'black-cone' hill. The low hills to the east were diverse in colour, ranging from dark red to olive, pink and black. These striking rock formations were to act as signposts for this section of the Sands when the rains came and altered the course of the wadi. Tracks spread out widely and there was no precise road. We simply headed south, with the sun beginning to set over the other side of the forests. Deep ruts in places indicated that the area could become quite muddy when it flooded. After 45 minutes, the 'black-cone' hill suddenly appeared out of the haze and we turned due west towards the forests down a track made by oil exploration teams to find a suitable campsite for the night. The prosopis trees formed a band of thick vegetation up to perhaps a kilometer wide in places. In the evening light, the woodland effect was out of place in this desert landscape but most welcoming. The track wound and twisted through the trees just as it might through an oak wood in England. We came across a field of dunes, each some 20 metres high, that must have been 2 kilometres across and 5 kilometres long, and skirted north to a clump of trees that was visible on the aerial photographs. Here we settled in for the night.

We made a fire for cooking with some dead wood just as the Bedu might while Kriss talked about the Sands and the Bedu and the affection he already had for them. He had been serving in Oman for several years and seemed to have experience in every conceivable outdoor activity. He probably would have been in the SAS had it not been for his imperfect eyesight. Here was action-man personified; he could fly, dive, parachute, ski, navigate and mend a broken vehicle with his Swiss army knife. Driving fast reconnaissance units in and out of the Wahiba Sands was very much his forte and indeed he was keen to share his new-found knowledge of the Sands, with which he had become intimately involved. Without Kriss's involvement, the project would have been a lesser affair.

He soon gained a reputation for saying yes to any request, and I suspect many on the team took advantage of his generosity. Wherever he went he carried field kit that would put any Boy Scout to shame. If you wanted a tool or instrument, map or dictionary, radio or flare, Kriss had it buried in one of his rucksacks and would appear with it moments later. 'You wanted a jeweller's screwdriver,' he would say with a smile.

Kit was very much a topic on the agenda as we discussed in every detail the equipment we might be able to borrow from the military stores. The list was endless and Kriss gave me invaluable advice on choosing quantities and specific equipment. As we went through his checklists it was quite apparent there was very little we would need from other sources other than the scientific equipment we were bringing over.

The next morning we rose early, having been awoken by a herd of goats and a lone camel scuffling around the camp. A thin layer of dew sparkled in the trees and on the grasses. My sleeping bag was sopping wet and I hung it over the vehicle door to dry in the anticipated sun; it had been a cold night and I had snuggled deep into the hollow I had scraped out of the sand.

We looked at the aerial photographs of the area and identified where we were in relation to the dunefield. It was relatively easy, as individual trees could be identified and our clearing was recognizable. We set off, passing the impressive field of white sand comprising barchans that Andrew had seen last January and considered suitable for his study to see how fast they moved in relation to different wind conditions. They were pristine. It seemed a sacrilege to be condemning them to such scrutiny, but knowledge of the geographical jigsaw requires intimacy; no corner should ever parry detailed inquiry. Pursuit of knowledge has a price; more usually this is the concern of life scientists or naturalists who must kill their recently discovered species to learn more of the animal or plant kingdom. In this case our footprints in the sand punctured the sharp outlines, to be repaired by the winds from the coast. Where sand and wind meet they will always create an infinite variety of patterns and shapes, now even more breathtaking in the early morning light.

As we rounded the southern edge the going became quite soft and our deflated tyres sank a few inches into the sand. Kriss, trying to be smart, got himself bogged down into a particularly deep bowl and for the first time we had to use the sand-ladders. After not a little swearing we moved on up to the ridge and on to a hill. Although not quite walking distance to the dunefield, from the point of view of access this would be most suitable as a field base. Kriss was already pacing out the mess tent, cooking area, laboratory and radio aerial. I took a series of photographs to be sent back to the geomorphologists in England. It was a perfect location for earth, life and human studies. All three scientific groups could be based here; convenient for the biologists wishing to study the prosopis and associated flora and fauna, the group surveying the dunes to track their movements, and the team who would be working with the Bedu to determine their lifestyles in and around these woodlands. A number of *barusti* shacks were dotted around and by the number of faecal pellets, goats and camels had left their mark in quantity. This woodland kingdom had a presence of its own and there was no hesitation that this should be one of the homes for the project in our quest to understand the Wahiba. We felt confident that it would serve us well as a research base both for the project and also for the Sultan Qaboos University if it was ever to follow up any of the Sands work.

Kriss then spent an hour or so finding the best routes in from the black cone as he was keen for a water bowser truck to be able to get there safely, both to fill up the water containers for the day-to-day running of the camp and to provide enough water for experiments to fool the plants into thinking the rains had come. Bedu tracks through the woods and a nearby well confirmed this to be a suitable spot. Pausing only to swig a coffee from Kriss's battered thermos flask we sped back to Mintirib, this time by the road that goes through the villages and on to the Ibra–Sur tarmac road. We drove too fast, but we did want to know how quickly we could do it in an emergency. This would be one of four supply routes for the project. On most expeditions the most dangerous places are dirt roads at twilight with ill-lit cars and lorries driving equally fast. Here there were added dangers from stray camels which would walk across the road regardless of any oncoming traffic, flashing of lights or hooting of the horn. Modernity had been too quick for their traditional pace of life and the highway code was not on their agenda. Mangled and bloody trucks crumpled on the side of the road, adjacent to a rotting carcass, was reminder enough to heed the dangers. We would be lucky if we didn't have a serious accident. 'Speed kills' was to become one of our mottoes but unfortunately most members of the team ignored it and continued to drive like demons.

When we got back to our headquarters we found that Taylorbase had already changed significantly, the huge, neoprene-covered hangars looking out of place in the desert and attracting the attention of our neighbours who lived in and around Mintirib. Word had, of course, been passed around from the Wali's office and already a number had asked if we wanted anyone to work for us. I had explained that our guides and helpers would be appointed through the Wali's office and that we had already secured the services of Said Jabber Hilays on the team. Salim, our immediate neighbour, had also made himself known to us. He owned a vegetable garden and would appear at all hours to ask if there were anything he could do. He was most helpful in letting us use his well until we had arranged for our water to be brought in by bowser.

Through Kriss and others we explained to Salim why we were settling in and to his query as to whether we would be there permanently, I said that we were in fact scientific nomads and that we would have to move on before the Sands got too hot. The camp would be taken down in April and was to be a temporary measure just for the winter. His eyes clearly expressed that he thought we were going over the top for such a temporary camp but he politely refrained from comment. Certainly Taylorbase was by far the most ambitious of the six camps I had operated on other continents.

Well-run base camps are the key to any expedition not continually on the move and should be encouraged for all scientific projects wanting to tackle integrated research. As John Hemming, Director of the RGS, often reminds us, 'Science requires time and not motion'. An effective base provides stability and security; it also becomes the soul of a project and its atmosphere will influence everyone who passes through its tentflaps. RGS base camps could be considered as extensions of the Society whose task is, after all, to offer a meeting place for interested parties.

I set off for Muscat, leaving Kriss to get on with 101 chores in Ibra including the submission of the final list of stores we would now like for both Field Base and Taylorbase. I stopped off to see Said and his family and he told me that he had been watching events at the camp with interest. We sat in his *majlis* (meeting room) for half an hour or so. He started asking me questions about the vehicles and tyres and whether I knew how difficult the Sands would be in parts. I tried to sound confident. I also explained that in due course we would also want to borrow some camels to travel to parts of the western areas where the Bedu were living in greatest numbers. Said looked incredulous. He asked me to repeat what I had said and then burst into one of his marvellous laughing fits, his whole body shaking with mirth and his headgear only just staying perched on his shaven head.

'But no one has travelled that far by camel for years!' he said. How could I explain to Said that we wanted to do this for personal gain?

Returning to the capital was like arriving in fairyland. The contrast was extreme. Colourful lights were strung everywhere for the fifteenth National Day celebrations on 18 November. Each year there are displays, pageants, military tattoos, fireworks and massed bands but this year they were to be the most spectacular ever organized. The lights, flashing furiously, covered some 70 kilometres of brightly lit motorway, with almost every house bedecked in some kind of patriotic pattern such as the crossed *khanjars*. Green, red and white, the colours of the Omani flag, predominated but the patterns were infinite.

The eight or so government ministries along the ministry mile were particularly resplendent. Each lamppost had a special decoration, ranging from turquoise-blue oil tanks rocking on a choppy sea to bright yellow butterflies flashing in the wind. Several million light bulbs had been specially brought in for the display and there was a small team permanently replacing them. Oman never does things by halves.

The heads of state that gathered together to witness a series of events, including a tattoo, the Sultan's speech and a firework display over the harbour in front of the palace, all went home with impressions that will last a lifetime. No expense was spared to make sure that the proud achievements and future visions of the Sultan and his people were shared with as wide an audience as possible.

I found it difficult not to go into a trance as I followed the hypnotically flashing lights. At first I didn't register when I saw a lone duo of bright blue lights, flashing out of time with the rest of the panoply, by my heartbeat quickened when I noticed a white police car now flashing me to pull over. I took off my earphones, turned off the Land Rover and leapt out to the first of several meetings with the Omani police. I shook hands and we exchanged greetings in Arabic.

'Do you normally drive around like that?' enquired one of the policemen, in perfect English. I stuttered and muttered and apologized, not quite sure what my offence had been. My civilian and military licences were checked, together with the vehicle, then the second policeman pointed out my error; my nearside rear light was broken. I pleaded ignorance and blamed a prosopis tree.

The Omani police were very much on the ball and we had been warned about driving desert vehicles in town. Dirty cars are forbidden by law and the dust that I had collected drew further comment. I knew I had had a close shave and shouldn't have let my high spirits carry me away so. I was much subdued as I pulled up in front of Capital Base to be welcomed by Heide and Chris Beal.

Capital Base was home for the project in Muscat. It became the epicentre of all town activities and coped with being a dormitory for all members passing through, a radio centre and an office for contact with all the ministries. It was John Cox who identified the need for such a base. I was originally against the idea as I knew that it would use up funds very quickly. But John had anticipated that there would be much more capital liaison than I had imagined. Without a central base information would become distorted and misunderstandings would proliferate. As word about the project slowly spread there would be a growing interest in the area and to some extent in our work, so a point of enquiry would be necessary. The Sands are a dangerous place but many Omanis and expatriates take a keen interest in the country's diverse habitats and this would be an opportunity to learn more about the great Wahiba.

I underestimated the clamour for involvement, not just by those who had come to know the project as sponsors and supporters but by those who had a genuine wish to visit the Sands. This was an aspect of the project to which I should have given greater thought. Let it be said that I had been advised by a SAF officer who had more vision than I that I needed a kind of VLO (Visitors Liaison Officer) who could manage and plan a programme of tours to the project sites. I was against this, thinking that we would turn into a circus if we were not careful. As it turned out, a ringmaster would have been a useful addition. There was to be a clash of conflicts and interests and those who wanted just to 'pop in' were soon to be an enormous burden on the project. The conflict was the real desire to share as widely as possible our work and our enthusiasm for the Sands balanced against the fact that there were only four months to complete the work. We wanted to have an open camp to make SAF personnel and sponsors particularly welcome but in the end we were to receive over 500 visitors to Taylorbase and problems did arise. It was Heide Beal who became the project's diplomat, coping with the turmoil and soothing fevered brows. In retrospect I do not know how we could have coped without Capital Base or Heide Beal. John Cox had been right.

One of the eight Racal radios remained at Capital Base and Heide would talk clearly to all sites within the Sands at 7.00 a.m. every morning; due to atmospherics, reception tailed off during the day and after 5.00 p.m. it was almost impossible. Heide was the link between the reality of the outside world and the sheltered life within the field. Situated within the Chelsea of Muscat, Medinat Qaboos, Capital Base comprised three houses lent by Taylor Woodrow-Towell which were part of the Palm Gardens complex, a set of comfortable air-conditioned family houses built around a beautiful bright-blue pool and a well-kept tropical garden. It was John Cox again who advised that the team should be whisked to the Sands before getting a real taste for the blue pool, the SAF Aqua Club and other capital attractions.

I was to spend over half my time on the project in Muscat. Often I would arrive late at night or early in the morning, dust-laden and field-worn, to be met by Heide's welcoming smile. Now, fresh from my encounter with the law, I was glad of a cheery greeting from Heide and Chris and we settled down to bring each other up to date with the progress being made either end. Having a phone put in, obtaining driving licences from the Royal Oman Police and passes for the Muaska Al Murtafaa Garrison, arranging a press reception when the team arrived and so on were all matters to sort out before the main phase began.

So the Sands were getting closer. That night I fell gratefully into clean sheets thoughtfully provided by Heide, my head whirling with everything I had to do in the morning as I dropped off to sleep.

Some two weeks later Heide and I followed the illuminated butterflies down to Seeb airport and dashed into the arrivals terminal for GF 004. It was to be the start of many visits to the airport as this group was just the first representatives of the various scientific disciplines to arrive. At 00.40 a.m. on Saturday 30 November Mike Holman and Nick Theakston pushed through the swing doors, their trolleys packed to the last inch. It was good to see the team – my partners on the administrative and operations group. Suddenly a burden was lifted; trying to be in three places at once had become a strain.

We went back to Capital Base to go through plans. Both Mike and Nick were in excellent spirits and raring to rush down to the Sands to get both Taylorbase and the vehicles ready before the surveyors arrived. A military driving test at 7.30 a.m. did not give them much time to combat jet lag and a full day sorting out vehicles, freight, insurance and some immediate shopping for stores meant a brisk agenda. We were due at Taylorbase on Sunday to meet Tony Hyde and open the camp. The surveyors were to arrive on Wednesday and our freight still hadn't turned up . . . we retired for a short night's sleep.

At 9.00 a.m. the next morning we had a working breakfast with Chris,

Heide and the project's liaison officer, Major Bill Davies, to delegate administrative priorities. We had already had a full day by then. Well before the sun was up, Mike and Nick were thrown into driving on the 'wrong' side of the road in an SAF Land Rover, overcoming a few difficulties with the Arabic road signs; had passed their driving tests, seen where our freight would be stored, been shown the camp laundry and the nurse's accommodation, and been introduced to the general layout of the MAM Garrison and the three messes, in one of which we now sat. As mess waiters in starched uniforms served scrambled eggs, toast and coffee, we cast an eye over the agenda and list of people to meet, which was of daunting proportions. We had to see representatives of the Ministries of the Interior, Education, and Heritage and Culture; pay a visit to the Meteorological Office, the university, an insurance office and the Natural History Museum; make contact with a journalist, attend a special concert – the list seemed to go on and on.

By the time we had decided on an order of priority we had finished our breakfast, thanked the mess staff and walked out into the heat of the sun. We had got a few things to do before we went down to the Sands the next day!

Tony Hyde had promised Taylorbase would be ready by 1 December and he had kept his word. Picking up Kriss and his *jundiis* (soldiers) from Ibra, we drove through the gates of the fort-like Taylorbase soon after midday. Tony Hyde took us round the finished camp. This was to be home for four months and it surpassed our wildest hopes. That it had been completed in just two weeks was a remarkable achievement even by standards in Oman, where new buildings go up at a speed unknown in Britain. Here was a base camp that would be the envy of any self-respecting field research station standing in isolation in a parched land. The rising wall of pink dunes could be seen 2 kilometres away across the wadi.

Plans for the project became a reality as we dropped our kitbags on the bare boards of the main accommodation block, staking our territories. At our disposal was full mains electricity, calor gas supply for the kitchen, four large water tanks, running hot and cold water, fuel dump facilities, comfortable bedrooms with 40 mattressed beds, full washing facilities for bodies and clothes, a camp mess with a kitchen, a dining room to seat 35 on one long table, a laboratory, a *majlis* (meeting room) for visitors, an air-conditioned sick room/surgery, a specially designed computer room with a double door to seal out dust and a large operations room with an adjoining radio room. Electric fans had been fitted in all the rooms.

The three hangars that made up the accommodation were constructed on concrete foundations and comprised a heavy-duty neoprene material slung over a metal framework – standard design for the construction industry. The rooms were built of softboard within this framework, providing continuous notice-board space. Although the walls were bare at present, it wouldn't take long to make ourselves at home. No base camp had ever had such facilities and we were all dumbstruck.

Two flagpoles framed the heavy wooden gates at the entrance to the camp, in front of which ten project Land Rovers would soon be parked. A *barusti* wall enclosed it, primarily to keep out the goats, windblown sand and, perhaps, the Wahiba bear that reputedly inhabited the desert.

Over the next few days Nick and Mike set to with gusto, co-ordinating the various comings and goings. Several tons of stores that had been borrowed from army warehouses and our own scientific equipment began arriving on Chris Beal's big transporters and were dumped in the central courtyard. A work party from Ibra came and made swift work of the pile, getting beds, lights, shelves, lamps, kitchen utensils, cooler boxes, route-marking pickets, jerrycans, camouflage nets and carpets into their respective places around the camp. Once the Land Rovers were delivered, Mike began setting up his toolshed and workshop and preparing the notes for the drivers which became known as the Rover Bibles.

Keeping a watchful eye on the goings on were many from the village, who were naturally perplexed at such activity. There was a growing barrier between the established pace of life in Mintirib and the rather frantic preparations for some 40 scientists. This would not have mattered if we had listened more closely to the villagers' queries as to why we were called the Wahiba Sands Project. It was Said who first warned me there was growing discussion relating to the naming of the project, for many of the tribal groups did not refer to the desert as the Wahiba Sands. This question had surfaced during the reconnaissance visits and had surprised us. Influenced by the maps of Oman, including those that had just been printed by the National Survey Authority, there was no doubt in our minds that the area had been called Wahiba for many generations.

'Just because the Al Wahibah live in the Sands,' explained Said one evening in his *majlis*, tucking into an enormous plate of rice and roast duck, 'you must not think that everyone in the Badiyah area considers it to be Wahibah lands. There are many who traditionally consider the Sands to be shared with other tribes and indeed it is often just called the Sands of the Sharqiya or Eastern Region.' It would have been naive of me not to think we would become involved in local politics, but the issues were lost on me in those first days. Little did I realize it at the time, but the project was to be used as a focus to resurrect old tribal conflicts. Of the many visitors we had in those early days, there was one particular Sheikh who discussed this issue.

His name was Sheikh Said Al Kindi, a noble figure with a white beard and colourful *masarra*. A businessman from Muscat, he had been visiting a friend, the head of the Al Hajiryin tribe in Hawiyah just a few kilometres down the track from our camp, and had called in to enquire if there were anything he could do to help. He spoke perfect English and I used the opportunity to tell him about our scientific mission to learn more about sand deserts and above all why there was all this activity just to collect information about plants and animals. It wasn't long before we got onto the question of the name of the Sands. By now a full set of new maps from the Government had been pinned to the wall from floor to ceiling. I could show Sheikh Said the many mentions of the Wahiba Sands on the maps and said that it would not be easy for us to use another name. Sheikh Said was clearly concerned and tried to discuss the possibility of our using a compromise name. Arrogantly, I would have nothing to do with his suggestions.

'The academic community', I stated, 'has Wahiba indelibly stamped in all its literature throughout the world and it will only be changed when the Government renames the Sands.' This was to prove an undiplomatic response, but I did heed Sheikh Said's words of advice – and more so in the future, as the project became further embroiled in local politics. He was to become a frequent visitor to Taylorbase.

The first team of the project to arrive was that of Captain Chris Dorman RE and his band of surveyors. Their task was to take a careful look at all the routes in the Sands, to mark those that would be used by the project and to make recommendations on how we should operate. This route-marking phase was set to last the three weeks before Christmas and would be done in conjunction with surveyors from the National Survey Authority. Everyone knew it was an important reconnaissance period and would dictate where in the Sands we could operate either by ourselves or in convoy.

The map coverage of the Sands is excellent and the new series that had recently been published had been at a scale of 1:100,000 and had covered the whole of Oman. The maps showed the main formations of the Sands very clearly; the main swales and ridges, the mobile dunes in the south east and the prosopis woodlands. A few wells were marked. It was our intention to survey the major routes for addition to the maps when they were next printed. In doing that, steel pickets 2 metres in length would be stuck deeply into the sand at kilometre intervals and marked accordingly. In theory this would provide a network of routes to avoid the team getting lost. It would also help the team work out their map references when reporting data. Much of the information collected would serve to corroborate what had been gleaned from aerial photographs. Confirming topographical features of any map is a continual process; in the Sands, where few have been before, much of the maps would be checked for the first time.

The surveyors were off at 6.30 a.m. the next morning under the leadership of Kriss and a convoy of CSF drivers and vehicles. The SAF had kindly laid on a Defender to fly some of the survey party up and down the main routes to see where they led to and to check if the Bedu were already crossing the Sands following the summer winds. Their task was to check four routes, M1, M2, M3 and M4. These routes would be marked with pickets every kilometre and would follow the course of the tracks both across the Sands in the middle and the two perimeter roads. In theory it would be difficult to get lost and a series of A and B roads would be added in due course. Operationally it would avoid confusion if a plea for help began with, 'I am at the 64 kilometre picket on the M2', rather than with a misinformed grid reference. This method would help those who have difficulty map reading; confusing the x axis (along the bottom of the map) or the y axis (along the side of the map) has caused chaotic rendezvous on many projects. Furthermore it was an opportunity to test the Land Rovers and the Magnavox satellite navigators that had now been fitted to them.

Satellites orbiting the earth emit signals which can be picked up by receivers on the ground. These little boxes, no more than 30 centimetres wide and a few centimetres high and fitting on to the dashboard of the Land Rovers, could 'lock-on' and 'track' a satellite as it went overhead. We knew when it had locked on, as it said so on the display panel. By knowing the position of the satellites (which have fixed orbits and circumnavigate the earth on a regular circuit), it is possible to calculate your position on the ground, so long as you have accurate time to the nearest second. These navigators could fix a point to an accuracy of at least 200 metres and in many cases much smaller. We had four and they generally aided travel in the Sands for all who used them. One of the more sophisticated versions, which had been connected to a compass and the vehicle's prop shaft, could determine where the vehicle was as it travelled along. It was therefore possible to punch in the co-ordinates of a known location such as a well or a particular prosopis tree and the navigator would give you a bearing to follow. This was mainly used by those who wanted to re-find locations as part of their monitoring work.

As the surveyors and their navigators headed off into the Sands I returned to Capital Base to help Heide. The drive through the Hajar mountains took about two hours as there was a great deal of traffic and I thought of the team in the Sands, where there is a very real temptation to drive around as one might on a large open beach. On this trip I witnessed a bad accident on the road not far from Fanjar where three people had been badly injured and most likely killed – a result of overtaking on a blind corner. I was to have nightmares about the drive and at one point had a near miss with a fast car. On my return to Taylorbase I learnt that two Land Rovers had received dents from driving too fast down slipfaces. This was a bad start to the project and my concern was expressed in a meeting we had when everyone had assembled. The three days we set aside to focus on such issues were very important for all new members of the project and would ultimately save lives.

Apart from a misunderstanding with Kriss about a team photograph to be taken by the *Daily Mail*, and an exchange of words about leaving the camp unattended one evening which was entirely my fault, the build-up to Christmas went without a hitch. We were now beginning to come to terms with the sheer size of the Sands and had a much better idea of where we were likely to be able to travel unhindered. By 19 December we were able to mark the routes on our OPS map for planning purposes and we had identified, both from satellite images and from ground observations, a huge area that comprised very difficult sand conditions with little evidence of Bedu living there. It was decided to make this a no-go area which we would visit only in convoys and possibly try to traverse from north to south. Mike had sorted out the vehicle plans, the radios were working well and Nick now ran Taylorbase as if he had been doing this kind of thing all his life. In the computer room, John Grannan, one of the few trained computer operators we could find who knew the practical problems of bringing computers into a desert environment, had set up six operational computers and was beginning to train the whole team in their use with enormous patience.

Using computers in the field is relatively new in the world of expeditions; I had telephoned a number of manufacturers and got some pretty curt replies. IBM stepped into the breach, however, and John went to see them to discuss our objectives and to be trained in computer surgery. It was his aim to have the whole team familiar with at least the word-processing and data-managing programs before flying back to England, and the time-honoured system of gold or red stars denoting progress against each member proved to be an effective method of encouragement. Even the most sceptical tackled the system with gusto and it is due to many hours of patience and good humour that every single member of the project completed their scientific reports on disc before leaving Oman.

Before we knew it, we were saying goodbye to some of the surveying team who had successfully finished their reports with information about all the routes in the Sands. On the night of their departure the project's banner, which was nailed to the outside fence of the camp, was cut down and removed. This banner had been printed in London for the launch of the project and said 'Oman Wahiba Sands Project' both in English and

Arabic. Only the corners remained. The discussions with Sheikh Said Al Kindi suddenly took on more importance and I wished I had heeded his warning; now there was real concern about the project and the safety of the team. Rumours that the camp would be burnt down spread quickly. It was a significant act and was definitely linked to the long-standing debate about the name of the Sands.

We had been naive. I had failed to consider the implications of being so dogmatic; certainly we should not have put our banner up outside in both languages. That can only have provoked those to whom this was a sensitive issue. We certainly did not want to cause a rumpus; as newcomers to the community we wanted to establish dialogue, not create further barriers. Taylorbase would be the focus for further discussion and I believed the sudden interest in the project on the part of the community would help our team to talk to many within that community about past and present-day life in and out of the Sands and what was envisaged for the future. Certainly the name issue was to become much discussed throughout the rest of the project right up to the time of writing.

Another cause for concern was whether the Wali of Badiyah would officially open the camp in front of the Omani press as had been arranged. We knew from discussions with the Wali and others that the matter would not remain quiet and that we might even have to change the name of the project. As an interim measure the Wali posted some night guards to make sure no more nocturnal visits would cause us harm. We always locked the main gates after the flags had been taken down at sunset and in due course we were to have two nightwatchmen permanently employed by the project, to do alternate days; they came from two different tribes namely Ali bin Abdullah bin Ha'atrush Al Wahibi and Salim bin Hamed bin Rashid Al Hagari.

Our Christmas break was shortlived; on 29 December we were at a spick and span Taylorbase waiting for two SAF helicopters to arrive from Muscat loaded with a press entourage for the official opening. Everyone had put on their Sunday best and some were even sporting ironed shirts and RGS ties. Mike Holman had earlier hunted around the *suuq* with Said to find a length of bright pink ribbon which we had draped across the gates for the Wali to cut. We were prepared for a major influx of guests as we had invited all senior members of the community and there was some apprehension concerning the growing discussion about the name of the project.

All new buildings need to be officially opened and it was appropriate that the Wali of Badiyah should do the honours. Before he arrived in his official Range Rover, a number of other dignitaries took their seats under a camouflage net just outside the gate. The Commanding Officer of the Desert Regiment and the Chief of Police from Ibra had come with a police escort and were taking their seats. Ralph Daly had arrived from Muscat with the press corps, who were now busy setting up their cameras. The Na'ib Wali (deputy) dressed in his finest, six local Sheikhs from the surrounding villages, some 20 government managers from the Mintirib schools, the post office, the hospital and the local sports club, and a number of others from the Mintirib community were now gathering in the makeshift theatre. Both Said and Sultan were there, Sultan having had to leave a recalcitrant racing camel in the early hours to be with us. It was good to see their smiling faces and I was reassured to know that they could help explain matters if the question of the name got out of hand. Said, his eyes twinkling, kept tapping me surreptitiously on the shins with his camel stick to let me know when I should move forward to welcome a new face or not. He understood my dilemma about protocol and that I had difficulty distinguishing Sheikhs from lesser mortals; hierarchies are a fine point so I welcomed Said's prompts. Everyone, without exception, had turned out in their finest ceremonial clothes; bright white *dishdashas* and polished *khanjars*, camel sticks and carefully tied headscarves gave the gathering a sense of occasion which, of course, it warranted.

Promptly at noon the Wali of Badiyah. Sheikh Ali bin Nasser, arrived in a ceremonial *dishdasha*. Said had given me a heavier whack on the lower shin as if to say 'this is it'. It was good to see the Wali again and he was obviously relaxed and not overtly concerned that we had our banner stolen. We went straight into a presentation, using the newly marked map of the Sands to show where we were planning to operate. The interpreter from the Ibra camp helped me translate into Arabic and I used the camel stick that Said had given me as a pointer. I explained why this was an important day for us – the day we opened our camp to study the Ramlat Al Wahiba, the Sands of Wahiba. I gave a thumbnail sketch of our interests in the Sands, how old they were and where they came from, the people, plants and animals that live in them, the grazing potential of the area and the night mists that maintain such a rich diversity of life. I went on to explain the ecosystem approach incorporating the earth, life and social components and how we were collecting the jigsaw pieces to try to get an overall picture of this unique area.

I told the assembled company that the team comprised scientists from Britain and Switzerland and representatives from the government ministries but that the most important members of the team were the local Bedu who live in the Sands. They were to be an integral part of our team and we were looking forward to meeting many more from the various tribes that live in or around the Sands, including the Al Wahibah, the Hajriyin, the Hikman, the Al 'Amr and the Al Bu 'Isa. I ended on the famous quote by His Majesty Sultan Qaboos bin Said about the value of natural resources and why it was necessary to collect facts about the environment before making irreversible development decisions. I was pleased to see there were nods of approval all round. Finally, before we broke up, I explained that it was an open camp and we looked forward to welcoming all members of the community to the camp and the *majlis* any time of the day or night.

We gathered round the pink ribbon across the camp gates. The Union Jack and the Omani flag flew proudly above. The Wali paused and said a few words welcoming us to the Badiyah area, extending the offer of any help and wishing us well with our studies with thanks to His Majesty Sultan Qaboos and praises to the Almighty and Most Merciful God. He then proceeded to cut the ribbon while cameras clicked. There was a round of applause and we followed the Wali to a line-up of seats in front of the camp notice board and took our places for the official photograph. A tour of the camp followed and then our own special *hafla* began, spread out in the courtyard on carpets and under yet more camouflage nets. Taking off our shoes at the edge of the carpets, we waited until all were present and then crouched down to be served dates and *qahwa*, which in turn were followed by eight very large trays of specially prepared goat and rice, 'special' from the kitchen. I caught John the cook's eye, and thanked him. I sat between the Wali and Ralph Daly and we agreed that the sight of everyone now communally tucking into such a delicious *hafla* meant that the camp was now well and truly operational.

This get-together had been important. Taylorbase had made contact with the local community both officially and informally and a number of potential misunderstandings had been cleared up, including the idea that we were oil prospectors looking for new locations to drill. The welcome we had received was encouraging. The people of the Sands are naturally hospitable but word would get out about our interest in the natural resources and I knew we would receive valuable contributions to understanding the Sands. We could now get on with the work.

CHAPTER FOUR

The Land of Sand – The Earth Scientists

Andrew Warren and his team studied this land of sand from an earth science standpoint, bringing together a variety of geomorphologists, surveyors and hydrologists to explain the diversity and shapes of this enormous, ever-changing sandpit. For most of the project they were based at their own camp on the east side of the Sands and it evolved its own life and character. It was a beautiful location on a hill in the middle of the prosopis woodlands with commanding views across the Sands to the west, the woodlands to the north and the mountains, including the 'black cone' overlooking Wadi Batha to the east.

Life at this camp, known initially as Field Base and later called by the local name 'Qarhat Mu'ammar', had a pace of its own, in contrast to the bustle of Taylorbase. The actual research sites where Andrew was to set up his dune-monitoring equipment was in fact a little distance from the camp. Travel to and from this was not always without incident and I recall one occasion involving surveyor Simon Kay and Andrew.

They had noticed the sun was nearly below the horizon and started to head back to Qarhat Mu'ammar. Their routine, established over two months, was to collect their surveying equipment, water bottles and day-sacks and climb into the Land Rover to take a track through the dunes to Field Base. But Simon, ever one for taking short-cuts, had tried a new route that morning and had had to be pulled out by a team of Kriss's Omani *jundiis* (soldiers), Simon does not like being beaten and wanted to show that he could get through the softest sand. Checking his tyres were down, he jumped into the driving seat with a determined look. Andrew, who was looking forward to getting back to his tent for a curry and chapatis after some ten hours on his dune, adopted a weary look while Simon gave life to the Land Rover.

By then we all knew that cruising dunes was enjoyable. With the right speed and determination the vehicle can roller-coaster over the sand, turning and twisting to find the best routes through and occasionally coming to a ridge that needs extra care and less speed. The traditional route curved south round the bottom of the dunes and on back to Field Base, but Simon determined not to be beaten, veered off the track to his short-cut, and fell well and truly in to the same hole.

Not to be outdone he tried every trick in the book. Getting Andrew to push and dig away the blocked sand, letting the tyres down further, using the sand-ladders, rocking the vehicle round on the balloon jack, swearing, asking the Almighty for help. The more they tried the more the sand gripped them. Eventually, there was not much room for manoeuvre as the Land Rover was wedged front and back between two slipfaces. It was getting dark. Defeat was admitted and for the first time Andrew was allowed to go to the emergency box in the back of the vehicle and do what he had always wanted to: let off a parachute flare.

Three kilometres away at the camp, quietly minding his own business, Clive Agnew, having just returned to his tent after a much needed wash, noticed out of the corner of his eye this little plea for help. He swore quietly to himself, but being a fellow lecturer from University College London leapt into his own vehicle and sped to the rescue. In the dark and amidst much cursing Clive managed to snatch-pull Simon's Land Rover out of the hole in the sands. The need for a second vehicle was demonstrated.

Back in the camp, Andrew reported his safe return to Nick at Taylor-base, who in turn informed him of the state of play elsewhere on the project. A lot was going on, and an eight-vehicle convoy was due from Muscat for the open-day at Field Base on the morrow, to be followed the day after by a large traditional goat *hafla*-lunch party with Sheikh Mohammed bin Hamad bin Daghmal Al Wahibi, the Chief of the Al Wahibah at Aflaj and his elders. A three-line-whip was ordered. Andrew just followed 'Fill-Base Out and Goodnight' with a few off-the-cuff thoughts on the state of life and went off to find the camp commandant Sayid Mbarrak to see if there was any supper left in the mess tent.

Very early the next morning, as he is wont, Ewan Anderson followed his customary route through the dunes to check his dew-collecting plates. He began at around 4.30 a.m. and today there was a very heavy mist. Following what he thought were well-tried tracks to the dunefield, he unknowingly locked into the new tracks made by Simon and before you could say 'Alhamdulilah' he too became bogged to the axles. Two hours later he was rescued by the Swiss, two elder statesmen from the life science team, Willie and Sonya Buttiker who were on the look out for dew-drinking beetles, biting ant-lions and other invertebrates. The earth scientists were not to follow Simon's route again.

Field Base or Qarhat Mu'ammar was the same place depending on to whom you were talking; situated right among the prosopis woodlands, it was home for those based on the eastern margin of the Sands near to the dunefield. This was the base Kriss and I had recced back in November; now established as a tented camp for the project run by the Coast Security Staff. It was the location that Andrew Warren was to make his domain. Andrew is a legendary figure in the world of sand and he was now in his element. He is one of the world's small band of desert geomorphologists and he believes passionately that to come to terms with desertification and sand movements the academic community must learn more about the dynamics of sand and under what conditions it moves – good, sensible stuff which in practice requires a great deal of detailed fieldwork.

We were lucky to have Andrew as our earth science leader; he is a true desert spokesman with a life-long quest to comprehend the Sands. He has a vision about managing deserts through a better understanding of how they work and, as Programme Director, he was never short of ideas, enthusiasm and commitment. 'How about if we . . .' was to become a catch phrase which we all mimicked from time to time, but ideas lead to discussion and discussion leads to action. Deserts and their inhabitants are in need of some.

Bribing the appropriate head of department at the universities to secure the secondment of geographers to join RGS projects is now commonplace and takes a variety of forms. With Andrew we had an advantage for Ron Cooke, the co-author with him of *Geomorphology in Deserts,* is his boss at University College London. This authoritative magnum opus is a well-thumbed reference for checking the terminology and complexities of sand dynamics.

But the pool of knowledge about deserts is not yet sufficient to advise prime ministers and presidents on how to stop desertification. This is not very encouraging for the 600 million people who live in arid regions, but since Brigadier Bagnold, that pioneer of desert studies, wrote *The Physics of Blown Sand and Desert Dunes* in 1941 there has been a slowly but steadily growing body of published knowledge on to which contemporary geomorphologists like Andrew are building. It is spiced with new jargon which even those in the know have difficulty in remembering. When is a barchan not a barchan and what happens if you cross it with a seif or is it sief or even sayf? As the Wahiba Sands have the greatest variety of dunes anywhere in the world, including good barchans and seifs, it is not hard to understand why Cooke and Warren consider this area to be of importance to desert research.

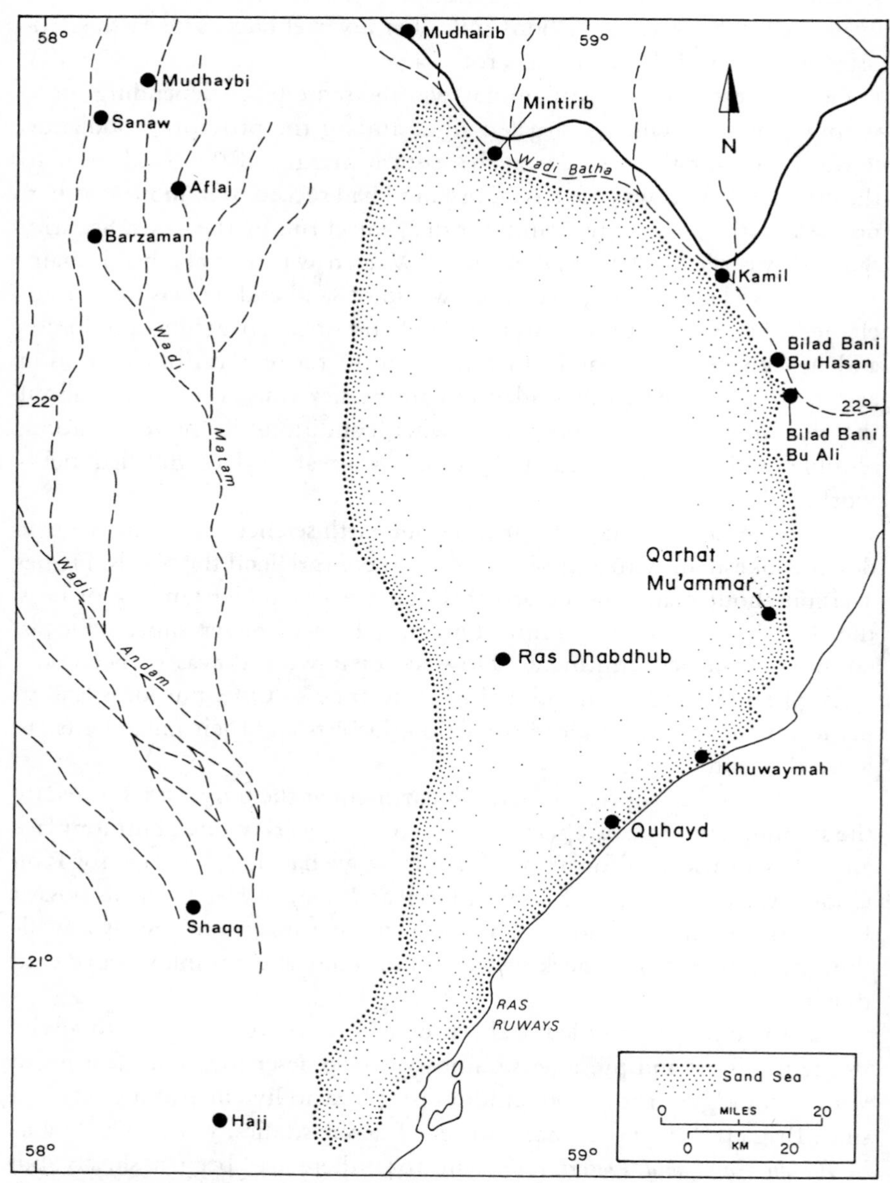

MAP 4.1 The Field Base was located at Qarhat Mu'ammar on the eastern side of the Sands.

Wind (aeolian) and water (fluvial) over the past several million years have been shifting and moulding these sands into the shape they are today. The sand, either 'fixed' to form semi-permanent hills or 'active' in moving dunes, holds the water upon which all plant life depends. People and their animals rely on the plants and so a proper study of the Wahiba Sands must consequently begin with the sand itself.

The Wahiba Sands are unique. They have been described as a perfect specimen of a sand sea (or erg, as they are called in North Africa) but they have been relatively little studied. It is only in the last eight years that the road through the Sharqiya to Sur, passing the north end of the Sands, has been open. While few scientists have visited the Sands, even fewer have ventured to cross them for fear of becoming bogged down in the soft patches. An expedition is the answer; the knowledge we collected would give others confidence to enter and go beyond the initial barrier of the high pink dunes. Being only two hours' drive from the new Sultan Qaboos University must now make the Wahiba Sands the most accessible sand desert in the world. They have the potential to become a living laboratory for international desert research.

According to Cooke and Warren, deserts cover one-third of Earth's land surface or 50 million square kilometres of the world. There are a number of definitions of a desert, but for our purposes we classify them as areas with very low rainfall – less than 150 millimetres each year. Discounting the cold deserts of the Arctic and Antarctic, the main deserts are in the Americas (in the south-west USA, Mexico and in Chile), Africa (Sahara, Kalahari, Namib), Arabia, India (Thar), Mongolia (Gobi) and central Australia. All of them are roughly the same distance, north or south, from the equator.

Continually changing local atmospheric conditions create long droughts and shorter wet phases in all deserts. Drought may last from a few to thousands of years. In Oman some people believe in a 10-year-cycle. There is also evidence that the monsoon winds from the equator 7,000–12,000 years ago carried much more rain to Oman. Rock paintings of elephants, crocodiles and pastoralists with huge herds of goats and camels in parts of Arabia and the Tassili N'Ajjer mountains of the Sahara are some of the evidence.

Deserts are not all sandy. There are mountainous deserts, bare plains and basins; in the basins are dry lake beds or sand seas. Deserts are fashioned into distinctive landscapes by wind and water.

The wind moves particles in three ways: suspended in the air, bounced along the surfaces or rolled along the ground (creep). For any of these processes to begin, the wind must attain a minimum speed which depends on the size of the particles. Sand is the material that is bounced; it can be defined as unconsolidated mineral particles ranging in diameter from 0.08–

2.00 millimetres. When it is being moved sand can act as a sculptor's knife, cutting into soil clods, plants, buildings, vehicles and paintwork. When the wind deposits its load, sand can bury plants and cover roads. The landforms the wind creates include individual pebbles (ventifracts), grooves and ridges in rock outcrops (yardangs) and, of course, dunes.

In those deserts where there are huge supplies of sand, there are sand seas, such as the Wahiba Sands. Other well-known sand seas include the Rub' Al Khali, the Grandes Ergs in Algeria, the deserts of western China and the three great sand deserts of Australia (the Great Sandy, the Gibson and the Simpson). Sand seas make up a third of the deserts of the world.

In each sand sea there are many different kinds of dune, but in the Wahiba Sands there are more than most. The classification and description of these dunes is an art in itself. Variety comes mainly from the complexity of the regional wind regimes, but there are many other contributors as well.

A simple but serviceable classification recognizes that dunes are either parallel or transverse to the winds, and that they range in size from ripples (wavelength 0.5–250 centimetres, amplitude 0.1–5 centimetres), through dunes (wavelength 3–600 metres, amplitude 0.1–100 metres) to mega-dunes (wavelength 300–5550 metres, amplitude 20–450 metres).

Furthermore, according to ideas that Andrew developed in Oman, dunes can be considered to have long or short memories depending on whether or not they can hold the results of a seasonal wind. The huge ridges of the central Wahiba Sands probably remember the winds of the last glacial period and are termed mega-memory dunes. Dunes only one or two metres high are obliterated by a strong new wind and therefore can be said to have short memories.

Water is the second force that shapes deserts. Any water that falls in the Sands sinks almost immediately and leaves no trace. On the mountains to the north of the Sands, however, it rushes down to the plains as flash floods. The water advances quickly on a fast flood wave (and woe betide anyone who has camped in the wadi that night), rises to a peak and then declines as it is lost by evaporation or soaks into the sands and gravels of the wadi floor. One of the major landforms associated with the floods is the alluvial fan, formed as they spread out at the foot of the mountains and on to the adjacent alluvial plains; the northern edge of the Sands is a great buttress of fans up against the mountains. Many of the wadis never reach the sea and spread out over low, flat basins, leaving a white salt crust known as *sabkha* when they evaporate.

Although one of the earth's smallest sand seas, the Wahiba Sands exhibit a very great number of dune types. Scientists see them as a huge laboratory for long-term research into dune processes but the Sands were to have yet more to offer, particularly to those interested in how they were formed

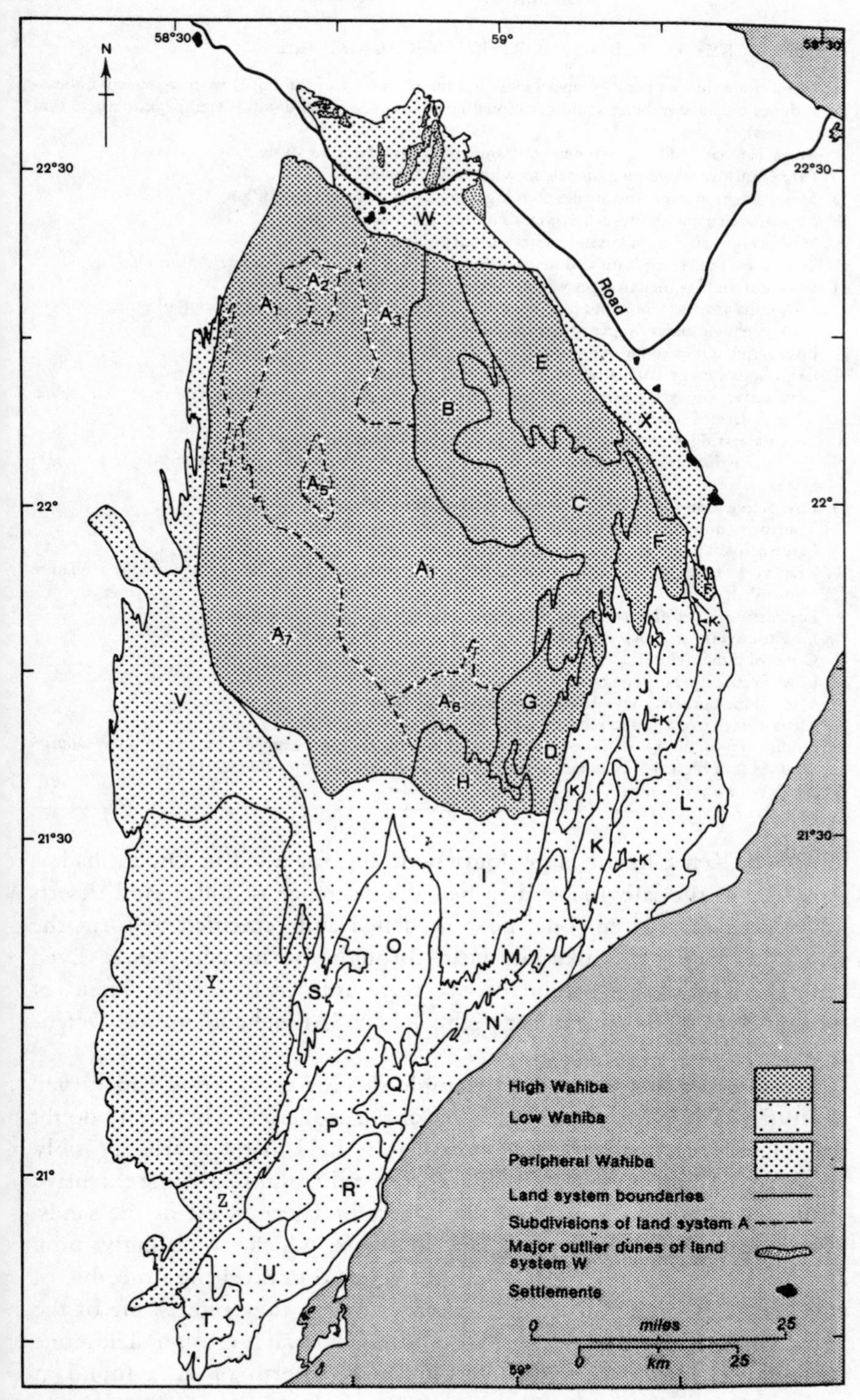

MAP 4.2 The primary landform units of the Sands. (Source: Oman Wahiba Sands Project Mapping Phase Report 1985.)

TABLE 4.1 KEY TO THE GEOMORPHOLOGICAL MAP – 4.2

A. Mega-dune ridges ('habl' – rope-dunes) and interdune swales (Shiqqah) with some active meso-dunes on sides and crests and occasional large transverse sayf dunes (Qarah) (3km apart, 1km across)
B. Mega-barchans and ridge-like mega-transverse dunes (Fuljs) in swales
C. Large but low active mega-barchans with Fulj in swales
D. Mega-ridges in risers and swales 20m high, 500m–1km apart, 5–8km long
E. Mega-barchans with mega-transversals on top and Fulj in swales
F. Mega-ridge-sized angular sand spears and major aeolianite area
G. Transverse dunes, medium and mega-ridges 20m high, 500–1000m apart, 5–8km long
H. Low, descending mega-ridges with overlying low mesa-dunes
I. Low mesa-ridges, vein dunes (irq), feather and thread dunes with an underlying NE mega-pattern all overlying older bigger cemented dunes
J. Low active dunes and low mesa-ridges 3m high, 50m apart
K. Large active mega-barchans with oblique NW element
L. Low active dunes 3m high, 50m apart plus long parallel low mega-barchans over 1km apart. Small area of aeolianite
M. Low mega-ridges, sayf-like dunes and large nebkha (qarhah)
N. Transverse dunes and barchans 10m high, 300m apart, some sayfs – no nebkha. Major aeolianite area
O. Low active dunes 3m high, 50m apart, small nebkha (niqda)
P. Transverse dunes and barchans < 10m high, 100m apart, N–S
Q. Large active mega-barchans also oblique NW element
R. Transverse dunes and barchans < 10m high, 200–500m apart, N–S. No nebkha, major aeolianite (dishshah area)
S. Low active dunes plus aeolianite in swales
T. Low transverse and crescent dunes
U. Crescent dunes (zibar)
V. Low mega-ridges capped by active or semi-active dunes
W. Mega-dune outliers, mega-barchans to east. Echo dunes
X. Mega-ridge-sized angular sand spears
Y. Dominated by blow-out dunes, dome shaped low mega-ridges capped by active and semi-active dunes 10m high, 1–3km apart. Major area of aeolianite outcrop and escarpments
Z. As Y, but no aeolianite

long ago. Ken Glennie, a geologist working for Shell in Oman, had as long ago as the late 1960s identified the existence of older sand deserts underlying the current one, now hardened and cemented to form the aeolianite rock that pokes its head through the moving sands. Even with this knowledge behind us, we were amazed to see the extent of these ancient sands during our flights in the SAF helicopters and Defenders.

In some exposed places, for example the coast, the wind had carved magnificent shapes of lions' heads out of the aeolianite. There is no doubt that the sands contain the largest aeolianite field in the world, possibly some two million years old. Furthermore, we also glimpsed some extensive examples of raised channels in the large alluvial fan west of the sands. These channels are the beds of ancient rivers, which, confusingly, now appear as long, winding ridges. If you want to unravel the timetable of past events, it is helpful to have raised channels; they offer some of the best records of the different periods of wind and water erosion. The stone tools which ancient men made from chert (a form of silica found in sedimentary rock) and left lying around the aeolianite and the raised channels allowed us a chance of contributing to the environmental history of the Sands; by finding how the Sands had developed in the past we might get a hint as to what might happen in the future. The first task of the earth scientists was agreed to be the reconstruction of the late Quaternary environment of the Sands concentrating on the three elements: raised channels, aeolianite and the evidence of early human occupation.

Working from aerial photographs, Andrew and his team divided the Sands up into a number of areas, classified by the dune patterns. Each was given a letter of the alphabet. The resultant map, with zones A–Z, was like a jigsaw puzzle. A more general classification divided the area into High Sands, Low Sands and Peripheral Sands.

The High Sands are the most striking, made up of long fingers of dunes, like a hand pointing northwards, each finger a large mega-dune ridge. These are the big dunes cut across by the 'pink wall' at the northern end of the Sands and up to 100 metres high and between 1.5 and 2 kilometres apart. They run parallel to the southerly monsoon or *kharif* winds, and Ken Glennie believes that they are a relic of the time when the winds were much stronger, during the last glacial period 15,000 years ago. They are also known as the *habl* which means rope, by the local Bedu, presumably because of the way smaller ridges of sand wind themselves obliquely round each of the mega-ridges in a spiralled, rope-like way. The big dunes are now fixed by vegetation.

The Low Sands are in the south and south east where there is a complex body of mobile and highly active dune networks, mainly of white sand and most of them aligned at right angles to the *kharif* winds, which blow southwesterly near the coast and swing southerly once inland. It is within this group that Andrew found a neat example of an active network for his studies on dune movement and in January 1985 stuck a stake in the ground for the research plot at Qarhat Mu'ammar.

The Peripheral Sands, to the south west of the Sands proper, is a zone within a thin sand mantle over a plain of eroded ancient aeolianite.

Andrew's team began by wanting to find out what had been happening to the Sands and their margins over the last two million years. What did they look like in an earlier age? Where did the sand come from and what clues had been left to show the wetter and drier periods? That was a tall order, so Andrew included three 'earth historians' or palaeo-environmentalists in his team. The first, Judith Maizels from Aberdeen, had had considerable fieldwork experience with expeditions to Finland, Greenland, Norway, Iceland and New Zealand – so much so, her department were loath to let her go. It was fortunate they did, for it was Judith alone who focused on the wadis, the river systems that provide the frame for the Sands.

With the exception of the Indian Ocean to the south east, it is the rivers that flow off the Hajar mountains that keep the Wahiba Sands in check and thus determine their characteristic scollop-like shape. The rivers only flow every three to four years and in the intervening periods the wadis are dry. Bordering the Sands to the west are the three overlapping wadi systems, Matam, Andam and Halfayn, fanning out over a large plain of gravel. To the north and east is the Wadi Batha, which passes Taylorbase and reaches the sea through two branches, the southern one near Field Base.

Based near Barzaman and armed with a geological hammer and aerial photographs, Judith combed the plain looking for raised channels. She located them with the air photographs amidst this scrub-covered land to the south of Mudhaybi. They are a spaghetti of quite extraordinary ridges up to 20 metres high, with pink mud and pebble cliffs where the wadis cut into them. Judith gained a reputation for being one of the toughest field workers, who brought her fresh and cheerful enthusiasm for her discoveries back with her dust-laden self to Taylorbase. As she worked on her own for much of the time, it was only through her regular radio calls that we knew she was well and making progress.

Judith was interested in any stone larger than 2 millimetres. These cobbles had been carried down from the Hajar mountains by ancient wadis. They provided a protective layer over the other wadi sediments and by resisting erosion when winds blew away the finer lighter material, they formed banks on either side of the ancient river courses. These banks can be clearly seen today as the raised channels, and their slopes and shapes are clues to ancient climates. The wetter it was, the greater the flow of the river; and the more meandering its course. From aerial photographs and ground observations Judith was able to determine five generations of channel formation – with a meandering oxbow river course being the oldest and straight channels being the most recent. These skeletons of ancient river are some of the best and largest in the world and provide palaeo-scientists with important data about the times when Oman was covered in forests.

During Judith's explorations she came across a new form of mud in one of the cliffs some 10 kilometres south west of Barzaman and so to add to her research she was able to contribute a new name to the literature: Barzamanite. She found that some of the sediments brought down by the rivers had now been chemically altered, possibly as water levels fluctuated in the soil and this had formed the pink calcerious clay which then baked solid to form the Barzamanite.

Rita Gardner complemented Judith by preferring particles less than 2 millimetres in size but, like the Barzamanite, fused together, this time to form aeolianite: sand blown on shore during the dry glacial periods when

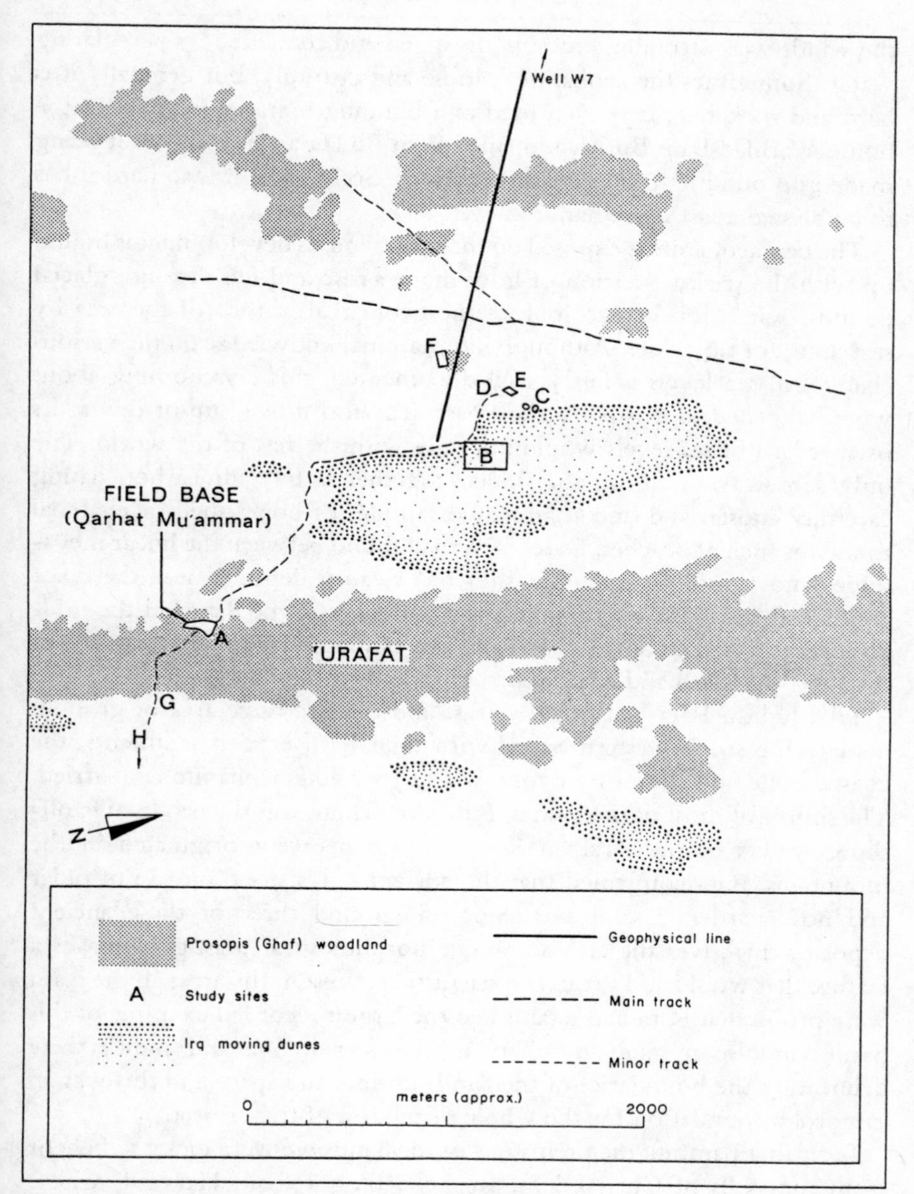

MAP 4.3 Study sites at Field Base (Qarhat Mu'ammar).

the winds were stronger and now dissolved and cemented by percolating water. Sometimes the aeolianite is loose and crumbly, but generally it is hard and rock-like. It is even used as a building material in some of the houses of Bilad Bani Bu Ali and Bilad Bani Bu Hasan and we saw it being made into building bricks by stonemasons. Sometimes it is so hard it has to be blasted apart by dynamite.

The best aeolianite is exposed on the coast, and its development is bound up with the tricky question of how the sea rose and fell through glacial periods. Rita's job was to look at the geological history of the area by searching for tiny clues. Although international knowledge on the various changes in sea levels is fairly well documented, this says nothing about what happened to this particular coast. If it also moved up or down, its own record of sea levels would not tally with the rest of the world. The only sure way of fitting the jigsaw together is by radiocarbon dating carefully chosen and uncontaminated samples of bone, shell or charcoal found on ancient beaches. Dates for a snail found between the linear mega-dunes and a mollusc found within old swamp deposits near the coast confirmed the existence of vegetation and lakes in the Sands in the mid-Holocene period 7,000 years ago and this ties in with similar reports from the Rub' Al Khali and other subtropical and tropical areas.

Like Judith, Rita had to move fast and cover a huge area of ground, namely the south-western Sands with their hard eroded sediments, the coastal zone and the eastern zone where rock-like aeolianite is quarried. The source of most of the sand in Rita's aeolianite was the productive off-shore waters of the Arabian Sea; only 10 per cent originated in the mountains. Rita confirmed that the current sands stood on top of older and now hardened sand seas. Since she found three of these ancient deposits, this gives the area some geomorphological prestige and was a feature that would add greatly to scientific interest in the area; all the team were proud that Rita had established the largest recorded example of this hard windblown sand anywhere in the world. She and Judith were delineating the boundaries of the Sands in time and space and this was an important foundation for the whole of the rest of the project.

In further painting the prehistory of the Sands we were lucky to benefit from the skills of Chris Edens, an archaeologist from Harvard, whose forte was an understanding of the stone tools made by early man in Arabia. These tools, or lithics, are made from specific types of rock found to the north of the Sands (cherts). The places where the rock was chipped to form tools such as hand axes, spearheads and arrowheads are known as 'factories' and Chris located any number of them. He also found a 'noise', or thin scattering of artefacts, throughout the Sands which showed that the whole area had been occupied several thousand years before. This complemented the work of both Judith and Rita because the tools, to his expert eyes, gave the date of their manufacture. Lithics are characteristic of the periods when they are made. The 'Arabian bifacial tradition' was particularly well represented in the Sands indicating a period of occupancy 7,000 years ago. This tied in very neatly with Rita's dating of shell fragments and pointed to the time of man's first exploitation of the area.

Chris found that the collections from one site in particular – Quhayd, on the coast – contained rather narrow pointed blades not found before in Oman. Because they had similarities with some that had been found in Dhofar to the south, he called the new arrowhead or spearhead the 'Quhayd variation of the Dhofar type'. This Quhayd variant gave Chris a clue to the kind of community living at the coast at that time. A lack of tools of other traditions suggested that the coastal settlements probably did not have significant contact with other parts of the world and were fishing communities. However, Chris was careful to emphasize that pastoralists tend to leave little archaeological evidence and any reconstruction of past occupancy of the Sands must be aware of the 'archaeological invisibility' of pastoral communities.

Although Rita had pinpointed the origins of her hardened sands in the rich waters of the Arabian Sea, where dead creatures falling to the sea floor had formed sediments which had then been taken to the beach by currents blown on shore by the strong winds, it does not mean that all the sands had arrived by the same system of 'erosion, transportation and deposition'. Discovering their origin was the task of Robert Allison, a research fellow at Durham University.

Bob wanted to ascertain the 'source areas and intermediate stores' that contribute to the Sands and to assess satellite imagery as a tool for identifying the mineralogy of the sand. To do this, he needed to complement his field observations with detailed laboratory analysis of samples by which he could determine their grain size, surface texture, shape and mineral content. This necessitated taking a large volume of sand back to Durham. His survey took him to over 400 locations, down tracks where few others ever went, along seismic lines, down most of the swales and finally down the out-of-bounds section on the east side.

The sediments in the High Sands vary from black to red in colour. The dark grains have been transported by the wadis from a group of rocks called ophiolites in the mountains, while the red grains are either broken fragments of cherts found in the mountains or particles that have been stained red by iron oxide. In the Low Sands and near the coast the sands are white or yellow, containing shell fragments from the sea. The marine organisms give the sands a high carbonate content and hence they are referred to as carbonate sands. Bob was concerned that his samples of sand might not reflect the thin veneer of material on the ground that was just a few grains or so thick as he believed that this top layer held the key to how the sand had moved. So, in taking his samples, he added a special method to collect only the surface grains.

It was a painstaking job to erode, transport and deposit his 400 samples all the way back to Durham. I watched Bob showing his method to a group of students from the Al Mutainabi school from Ibra. At each location, a point was identified as being representative of the overall site conditions. First, using the hands, sand was bulk sampled from the ground surface at 90° to the direction of the surface ripples. The top 1.5–2 centimetres of sediment was taken over a 10 × 10 centimetre area. This was sealed in a plastic bag. Damp sand was dried before final packing.

The second stage needed two sets of hands. A vegetation-free site was chosen at least 20 metres away from the vehicle. The students used an adhesive-backed transparent plastic sheet one square metre in size. The backing paper was carefully peeled away and the upper surface wiped with a cotton cloth to remove any static charge; this was to avoid sand grains jumping up towards the sheet. The sheet was now taken between four hands and lowered to the ground surface, held at the corners and gently massaged to make it adhere to the sand grains beneath. The orientation of the sheet was marked using a Silva compass. Once the plastic had picked up the topmost layer of sand it was replaced on the backing paper, folded and stored. By repeating the sampling, Bob could work through a particular site until the top 2 millimetres had been sampled overall. You have to be keen to do that 400 times over an area the size of Wales. However, 15,000 square kilometres, in places over 100 metres deep, is a lot of sand and Bob's tiny sample of this huge bulk will give the first hints as to where it has all come from.

Because so much of the Sands came from the floor of the ocean, Bob decided to look at this zone, too, with the help of the Sultan of Oman's navy and a large powdered milk tin with holes in it to let the water out. Kriss and three navy divers helped him sample nine transect lines between Ras Ruways and Khuwaymah, using this scoop. Positions in the sea were fixed by placing a Land Rover on the shore and fixing its position using a sextant. The ocean bed was confirmed to be covered with sandy material which appeared to become increasingly fine offshore and to the north.

Bob's sampling has shown that the sediment types within the Sands are numerous but, by identifying the composition of his samples, he can show the links between the two main sources of sand – the mountains and the sea – and the intermediary sources such as the alluvial fans and the wadis. He has been able to estimate the movement of the sediments within the Wahiba Sands and thus offer an explanation for the myriad colours for which they are famous.

Clive Agnew, from University College London, is an authority on moisture in sand and he joined the team soon after the rains; in fact, we were prevented from getting to the Sands on the day I collected him from Seeb because three of the wadis near Ibra had been flowing fiercely and cars were being washed downriver!

Working from Qarhat Mu'ammar, he had six weeks to tell us what happens to water in the Sands. When rain does fall a number of processes are put into motion and it is vital to understand them in order to know how to turn deserts green. Clive came armed with a number of special tools: an AWS (automatic weather station), a neutron probe, psychrometers and even a bowser.

Clive was popular for bringing the latter to Field Base. On loan from SAF, this huge tanker was filled from one of the nearby wells and used for Clive's false rain experiments. By having so much water to hand he was able to attain a high level of infiltration, while the rest of the team soon discovered it was ideal for a midday shower. Understanding the 'water balance' involves estimating changes in soil moisture. The equivalent to the 'erode, transport, deposit' (ETD) process for hydrologists is the 'infiltration, drainage and evaporation' (IDE) of water and these processes must be evaluated for different rainfalls and temperature conditions. The pieces of the jigsaw that hydrologists like Clive collect are considered to be vital for predicting whether a specific sand dune can or cannot sustain life and what the thresholds are, both in how much water is needed and upon what part of that dune it is best to plant, since some areas of a dune hold the water better than others.

The automatic weather station had been loaned to us by the Natural Environment Research Council (NERC). Clive set it up in the dune site. This sophisticated machine had sensors that record on tape the solar radiation, wind speed, wind direction, temperature and wet bulb depression (and therefore humidity) every five minutes for as long as the batteries were connected. Goats and a stray camel tried to interfere from time to time even though a roll of barbed wire encircled the contraption, which was in position for over a year. The digitized tapes from the AWS were sent to the Institute of Hydrology in Wallingford which collated the data and printed out hourly summaries. This was a valuable tool for localized monitoring as meteorological conditions vary considerably within the Sands and existing data from Masirah, Sur and Yalooni could only supply a fraction of the story.

There are large spaces between sand grains, making it difficult for water to be trapped within the pores as it is, for example, in soils and clays. In the same way that water will drain through the 'pores' of a kitchen sieve (the larger the mesh of the sieve the less water retained) so the relatively large spaces make it difficult for sand to hold water. Infiltration into sands is very fast except where plants are established, particularly on the vegetated mounds called *niqdas*. Clive estimated that the equivalent annual rainfall of London (600 millimetres) would have to fall on the Sands in an hour before they became saturated and water flowed on the surface.

In the dunes, the rate of drainage varies between the dunes themselves and the swales. Sands in general are only capable of retaining water equivalent to 5 per cent of their volume. This is known as their 'field capacity'. Following infiltration, the wetting front moves through the soil until drainage ceases, when the forces of gravity and the suction exerted by the soil matrix (the sieve effect) equal out. Clive showed that in a dune crest little water reached deeper than 120 centimetres and any excess water flowed sideways down inside the dune. In the swales water soaked down beyond 160 centimetres and was still moving down 14 days after the irrigation. Furthermore, Clive demonstrated that the areas under certain plants like prosopis were very much slower at taking up this water when saturated, but held on to a much higher percentage of it after a week or so. This knowledge of the sponge effect by the soil beneath plants would be valuable to the team of biologists.

Actual evaporation is very low in deserts simply because there is little water. Clive used evaporating pans near his research sites to measure potential evaporation and, of course, when there is water such as he was supplying, evaporation is very high – sometimes as much as 15 millimetres a day in March. Even with this high potential rate of loss Clive found the Sands were still moist some 40 days after the heavy rains in February, especially in the swales.

During the project Clive had to be wary of the attentions of local wildlife. While the fences around his experimental stations kept out itinerant goats and camels, smaller animals became adept at getting through the rolled barbed-wire fences and telltale pawprints indicated that foxes had drunk from the pan, so increasing the evaporation rates!

One of the distinctive features of our camp at Qarhat Mu'ammar was that it never went to bed; all night there were activities of some kind as one team or another continued their research. The flaps of one of the army tents would be flung apart at 4.00 a.m. precisely every morning by Ewan Anderson, setting off to collect dew before it all evaporated.

On certain nights the sands would brew up large quantities of night mist and deposit this in the form of dew on all suitable surfaces, especially sleeping bags. During the project there were a number of sightings of quite dense mists and once even a thick fog which rose from ground level to a height of 10 metres. At its thickest point, visibility was down to 100 metres and small sand dunes were only visible by their peaks. All the vegetation in the area was saturated and it was calculated that the mist was some 20 metres deep in the swale.

We were lucky to have Ewan with us to look at the role the dew left by the mists played in maintaining life in the Sands as this is possibly one of the least known and yet most important aspects of the ecosystem of the Wahiba Sands. Ewan, a much-travelled and highly experienced scientist, had been working for the Public Authority of Water Resources in Oman which had reluctantly parted with him for a period so that he could work from Field Base. He has a particular trait useful to expeditionary science, apart from not needing much sleep, and that is that he finishes his field reports almost before the tent doors have stopped flapping.

Ewan arrived at Qarhat Mu'ammar armed to the teeth with special gadgets to quantify this famous Wahiba dew, some more technological than others. In simple terms, he wanted to know how much dew would be landing on our sleeping bags, what conditions influenced this, what height above the ground it was to be found and where on a dune or near a prosopis tree you would get the highest readings. However, he couldn't use sleeping bags as collectors as although they are effective, they do not represent the surface of leaves so he tried to find a surface that had the same 'emissivity' or 'attraction'. His most popular sensors, little metal plates 55 × 75 millimetres in size, were coated with three layers of latex-based paint, because he found that this approximated best to that of a prosopis leaf. As a back-up, he also used copper-covered sensors. All the sensors were clipped in a variety of positions both on the trees and on metal rods on the ground, some near the AWS to relate to other meteorological readings. If any dew was collected Ewan would transfer the sensor straight into a clip-top watertight plastic container and weigh it in the field laboratory, using an accurate balance that could weigh to 0.001 of a gram.

Ewan scattered his sensors round Field Base, but had to limit the number to those he could collect himself in just five hours each morning. The last of his sites was by the coast so he could have an early morning swim in time to return to the camp for dhal and chapatis with the soldiers at 9.00 a.m.

Although his special sensors were the main sampling methods, he also used some novel methods: mopping up with blotting paper impregnated with a cobalt compound; measuring the conductivity across gold-plated circuitry on similar-sized sensors (the wetter the palette the higher the conductivity); and taking small sprays from the trees and weighing them before and after the dew had been blotted off. Using a 'speedy moisture tester' that relied on collected moisture to react with calcium carbide and then measure the resultant gas (acetylene), Ewan was also able to measure to within 0.2 per cent the moisture of samples from the top layer of sand only 1–3 millimetres thick.

Ewan's results confirm that dew has a considerable effect on the life of the Sands; indeed, many animals, from dew-drinking beetles to the Arabian oryx, rely heavily on the dew for survival. For dew to fall the ground needs to cool and so the best conditions are clear skies, 75 per cent or higher relative humidity at sunset and a constant 1–3 metres per second windspeed throughout the night. When conditions were perfect individual sensors recorded more than 2 grams, a figure that would give the equivalent to 0.5 millimetres of rain. As the whole rainfall for the area is less than 10 millimetres per year this figure is significant. Furthermore, Ewan found that the prosopis tree he sampled just outside his tent absorbed the night mists through its leaves – a process that was of interest to Kevin Brown, who made a detailed study of the woodlands as a whole.

Ewan is a good example of a geographer who had considerable research experience. Being in touch with the field, its processes and its peoples gives him a broad knowledge of the world. Unlike many who have the heavy burden of academia, Ewan is a born communicator and accounts of his days at Field Base and his own studies in Oman will be topics of the many lectures he gives all round Britain each year and will no doubt inspire others to get up at 4.00 a.m. to see desert sunrises.

Andrew Warren clocked up more days in the sand than any member of the team. Most days, drivers permitting, he would be on site soon after 7.00 a.m. He had got to know his dunes intimately and built up a comprehensive picture of how his dune network behaved. It was complex. In the winter, the crests of the dunes, but only the crests, sometimes moved more than a metre in a day and often buried survey pegs, but the underlying dune itself remained stable. In the summer things are quite different. Whole dune ridges move 15 metres forward and change their shapes entirely.

Andrew is an advocate of understanding the dynamics of sand movement under different wind conditions as he believed that dune control would be much more effective if the mobile zones within dunes were identified. To get an accurate picture of dune movements there had to be an accurate survey.

Andrew had three sites. The first was close to the coast at Quhayd, near Mobile Base 3 – a dune 80 metres high, thought to be moving very slowly, comprised predominantly of coarse sands with prominent traces of dark red chert. Thanks to a small team of surveyors from the Sultan of Oman's Artillery, this huge dune, over a kilometre long, was surveyed within a week.

The second site, the smallest at only 100 × 50 metres, was at Ras Dhabdhub, at the northern edge of the 'hand' of mobile dunes reaching in from the coast into the Sands. The dunes there were only 1–2 metres high and the sand again contains red chert grains.

The main site was the 200 × 200 metre plot near Qarhat Mu'ammar.

Clive's AWS was nearby, so all measurements of how the dunes moved could be correlated to accurate wind measurements. All the survey data from these plots was first computed in the field using special survey software on an Epson lap-top computer and was then processed by the computer at the Department of Geography at University College London. This constructed Digital Terrain Models of each site to help analyse the dynamics of the dunes.

Dune dynamics are interesting. In a network dune, such as is found near Qarhat Mu'ammar, the sands are highly mobile. Even a slight breeze will start the crest moving. The stronger the wind, the more the dune is disturbed and so the further the crest moves. In the winter only the uppermost part of the dune moves; in summer the whole thing shifts bodily forward. Each wind from a new direction sets up its own pattern of ridges and slipfaces, superimposed on earlier dunes and, should a third wind come along, then yet another pattern is superimposed.

The wind that blows most strongly and consistently maintains a dominant dune ridge. No succeeding gentler wind can modify it. Different annual regimes do occur from year to year and place to place, which means you may get a different pattern each year and at each location.

There are yet more complexities. When a transverse ridge crosses another transverse ridge a node forms. The bigger the two ridges, the larger the node. Furthermore, a second set of 'interference' linear dunes can be superimposed on top of the network system.

There is one further twist. The studies during the winter of 1985–6 showed little movement of the dunes as a whole but just their crests. However, between March and July 1986 the strong *kharif* that caused a major shift north eastward of the main slipfaces also brought about a distinct change in the overall shape. The hollows become even deeper and the nodes higher.

The task Andrew set himself was enormous. The project has but scratched the surface of the complexities, but he has been able to use the skills of other disciplines to help him get to grips with this particular dune network. Clive, for instance, has shown him that the moisture in the dunes following last year's rains did not seem to have a major effect on their shape. Andrew has laid down a foundation, the first of its kind, that can now be built on by subsequent researchers through the Sultan Qaboos University.

By their research the earth scientists were able to uncover some of the underpinnings of this remarkable area and, in so doing, help both the biology and the social science teams to come to terms with its sheer magnificence and complexity.

CHAPTER FIVE

The Life Science Team

The pigeonholing of scientific disciplines is as marked with biologists as with any other group, aided perhaps by taxonomy and the gulf between those who study animals and those who prefer plants. Ecologists, who must understand the links between all living organisms, are beginning to break this mould but there are very few who have studied arid areas and even fewer who have touched the sand of the Arabian deserts. Consequently, the task of finding someone suitable to head the life science team was difficult. It was important to dovetail research with those who understood the geomorphological processes as well as those who could talk to the Bedu. Also, the diversity and richness of the biological resources of the Sands needed a spokesman who could describe the whole range of flora and fauna from the shortlived ephemerals to the mighty prosopis and from the moths that live off the tears of donkeys to the timid gazelle.

Such people are rare. One day there will be many environmental diplomats who will be called to advise prime ministers and presidents but meanwhile we have to seek out those who can already speak the right languages. If you want to represent an Arabian desert ecosystem you must first prove you can be sensitive to the complexities of the ecological fabric. Paul Munton with his shaggy beard and bearlike presence has this ability and is the only biologist I know who could take on the task we had set ourselves. A man of enormous presence, not given to unnecessary chatter, Paul had a gentle but resolute approach that was to balance the briskness of Andrew Warren and add to the strength of the team.

When we met in the Ennismore Arms near the Oman Embassy in London in November 1984 to discuss his views of the project I was immediately impressed by Paul's broad knowledge of conservation issues, his desire to understand the differences between the various zones of the Sands, his passion for Oman and his strong belief that it is one of the few countries in the world capable of integrating realistic conservation and development policies, and his long-term vision of being a spokesman for the natural resources of areas such as the Wahiba Sands. He is a consultant to IUCN (International Union for the Conservation of Nature and Natural Resources), in particular to their Species Survival Commission (SSC). Since completing his doctorate at Bristol University Paul has also worked for the Sultanate of Oman, the Nature Conservancy Council, the King Abdul Aziz University in Saudi Arabia and the World Wildlife Fund. He told me about the enjoyable two years he had spent studying the Arabian tahr, a rare goat only found high in the mountains of Oman; his findings led to the setting up of the country's first conservation area in the Wadi Serin of the Jabal Aswad, and the first guardpost manned by local people, the tahr guards. Through Ralph Daly's Office this led to a second very successful conservation project; Mark Stanley-Price's White Oryx Project at Yalooni. As we talked, packed closely together amidst the bustle of the midday drinkers, his bearlike laugh could be heard above the general shindig; through the haze, I could sense Paul was already back in Oman.

Two months later we conducted a low-level aerial survey of the whole of the Sands. Paul was in his element as he recorded gazelle, camels, goats and sheep, *barusti* settlements and the prosopis woodlands. The next day we drove with Kriss and his *jundiis* through the prosopis woodlands on the eastern edge of the Sands. Extending for over 80 kilometres, the wood was so dense in parts that the convoy of vehicles had to thread carefully between the trees and on a couple of occasions the sand-ladders on the side of the Land Rovers had to be removed to avoid damaging the bark. In spite of the three-year drought these trees, many over 20 metres high, looked only a little the worse for wear. We camped that night near the trees and, as one might find in any woodland, there was a degree of activity from the birds, lizards, flies and other animals as the sun set.

Paul was very much at home in this wilderness. He explained that the ten or so different plant species found in and around the Sands so far were only the perennials and that when the rains came that number would swiftly multiply. We were certainly going to need a botanist who knew the Arabian flora. Paul's collections would go to the Natural History Museum in Muscat and thence to the Royal Botanic Gardens, Edinburgh.

That visit to the Sands in January 1985 gave Paul a good idea of the team he would need to choose. He was able to map 14 zones in the Sands representing the main vegetation communities and to collect sufficient information to consider how knowledge about the biological resources of the sands would dovetail into the work of the earth scientists, the social geographers and the Bedu.

All around the world animals and plants have been named and placed in specific pigeonholes. In too many university departments and conservation bodies there is too little flow of information between the specialists. This situation is improving but by and large the person identifying mosquitoes doesn't necessarily communicate with the expert on gazelles and more significantly he doesn't think it important to do so. Most of the time it doesn't matter, but when you want to cast a wide collecting-net across an Arabian desert as we did the key question is how many representatives to have. Either you call in all those experts whose specific knowledge is indispensable or you pin your hopes on an experienced naturalist who can collect across the board and then send the various genera to the respective authorities, usually dotted all over the world. This is a dilemma for any research team, whether it be from the United Nations or a local university, which must survey a region in a relatively short time and make accurate statements about its natural resources.

Taxonomic information is the building brick of any environmental study but the taxonomy of plants and animals in Oman is extremely specialized and experts worldwide number no more than 50 or so. As we had promised to keep numbers to below 30 for the whole project, Paul's team were hand picked for a mixture of taxonomic skills and their ability to interpret the biological resources of the Sands. One of the specialists we were looking for had recently been promoted to Senior Scientific Officer of the Royal Botanic Gardens at Kew and appointed co-editor of *The Flora of Arabia*. Kew's interest in arid zones has long been recognized and applauded. At the Kew International Conference on Economic Plants for Arid Lands in July 1984 we were able to hear of the vast amount of research and fieldwork being undertaken to study the plants of arid regions because of their economic value. It was there we heard of the value of the *Prosopis cineraria,* found on the eastern margins of the Wahiba Sands, and it was there that we met Tom Cope. Tom is a taxonomist through and through. Taxonomists are some of the most thorough scientists in the world; when it comes to classifying a new plant or animal, the buck stops with them. Their skills are called upon by a wide range of authorities wishing to have plants checked or named for reports and publications, popular or otherwise. Tom had a long list of new taxa, scientific papers and publications to his name but his real interest lay with grasses and he was currently working on an analysis of the chorology (science of geographical distribution) of the world's grasses for a major symposium in Washington. As co-editor of *The Flora of Arabia* Tom was particularly interested in our Sands.

Taxonomists can choose the names they give to new plants. Tom has a plant, *Dimeria copean* named after him. I asked him if we could name any new species we found after one of our Corporate Patrons; in theory there is no reason why not, but there is an unwritten law for taxonomists, that the person whose name is included should have had some connection with the plant concerned. I thought this a shame as I would have liked to add that to the list of advantages I could offer some of the more generous sponsors. Field research is desperately short of funds and I feel that academia must allow some corporate involvement.

Tom, together with Michael Gallagher of the Natural History Museum in Muscat, proceeded to redefine the flora of the Sands. Tom had a nose for new plants and during his six weeks with the project found them in all 14 vegetation zones within the Sands, particularly in prosopis woodland, even though the rains hadn't come to engender the growth of annuals by the time he boarded the plane back to Kew. Mike Gallagher returned to the western fringe of the Sands after the spring rains and found a further eight new plants, making a total of 176, far in excess of what we had hoped for. They gave confidence to all who believe that all deserts one day can be turned green.

Kevin Brown is another natural field person. He was brought up in south-east Asia where his father had been an army survival specialist and where he had spent many holidays living in the forests. Before the project he was with the British Trust for Conservation Volunteers, having qualified at Durham as a biologist. Keen to study for an MA and with an empathy with the forested world, Kevin joined us to look at the remarkable prosopis tree in detail. We knew this tree to be a valuable resource to the Bedu but we were all astounded by how much it was used and how it provided a valuable habitat, not only for the people but for the many animals and plants which had made the Sands their home.

This prosopis tree is one of the leguminosae which is a large group of plants that has been very successful in deserts; they have the ability to harbour nitrogen-fixing bacteria in their roots. These bacteria combine gaseous nitrogen in the soil with other elements to produce nitrogen compounds that can be used as a fertilizer by the plant. This is especially important in deserts where the soil is almost always poor in nitrogen. All plants in this group produce seeds in a pod (the legume). Many leguminous trees (the best known are the Acacias), are important in providing leaves and legumes as forage for wild animals and domestic livestock in deserts. The genus *Prosopis* is distributed worldwide, representatives growing in the harshest environments. *Prosopis cineraria* is found only in Iran, Afghanistan and the Arabian peninsula. It shows a lot of variation in shape and structure: it can be a substantial oak-like tree up to 9 metres high or a bush only 1 metre high. This variation results from its capacity for vegetative regeneration, which enables it to survive severe cutting by man

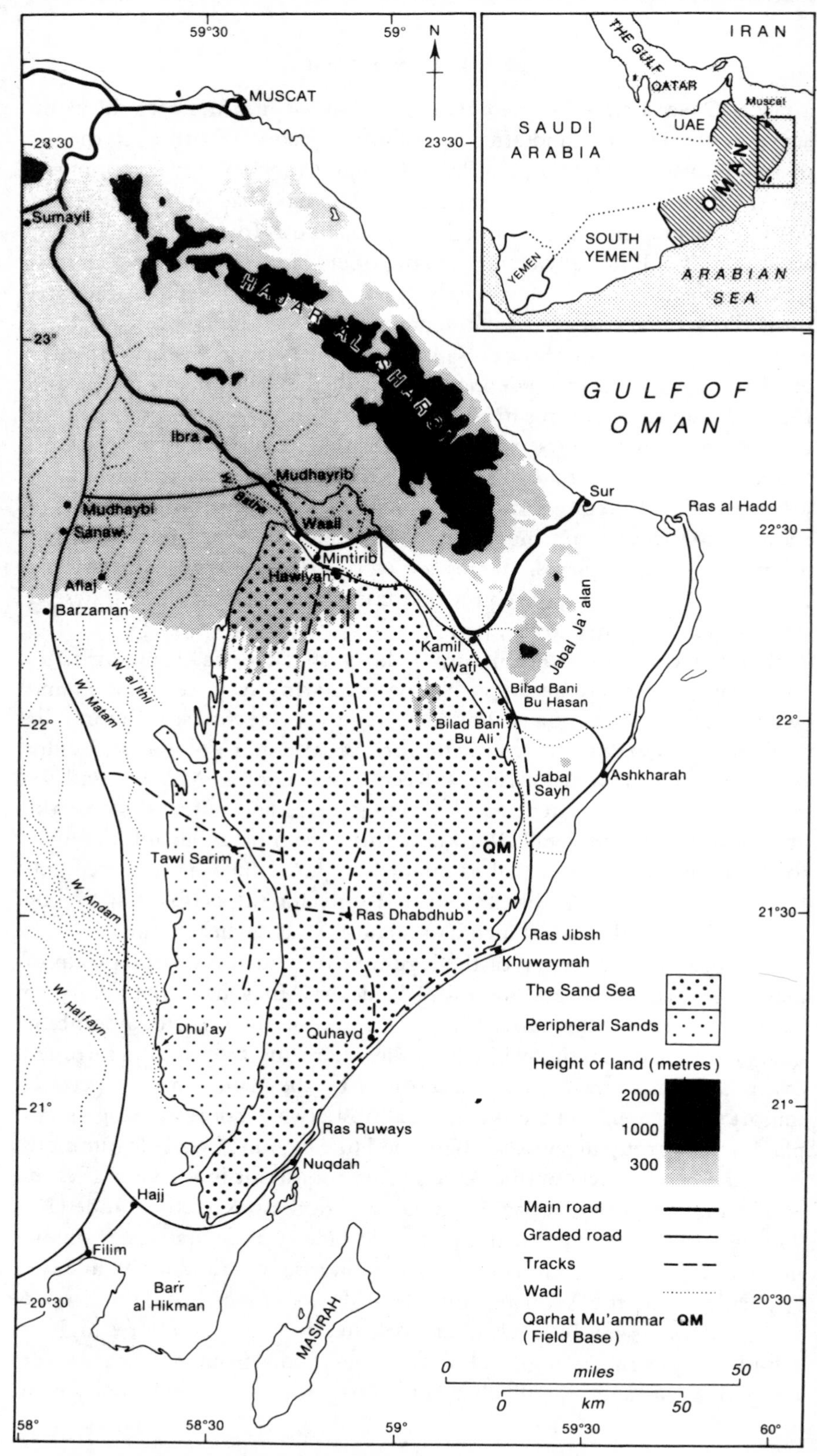

Map 5.1 Location map of the Wahiba Sands in the Sultanate of Oman.

and to grow out of accumulating and shifting sand. Paradoxically it grows rather slowly so the vigour of the species is expressed through its tenacity of life rather than fast growth. It has a thick, fissured grey bark and a heavy hard wood of high calorific value which is valued as firewood and as building material.

Largely based at Qarhat Mu'ammar, right in the middle of the woodlands, Kevin checked on his tagged trees and research plots with the aid of a Magnavox 6102 satellite navigator, which he had programmed to take him back to the same sites over a 70 kilometre range. There are three areas of prosopis woodland in the sands: a south-western area of variable density, a central sparse woodland and a north-eastern woodland which is the densest of the three. Running north-south for 85 kilometres and 20 kilometres at its widest point, the north-eastern woodland hugs the eastern edge of the Sands proper. Before the rains came in February the area was experiencing a severe drought, yet the trees were still green and the occasional one had small yellow flowers. So, while all the other plants and animals go into a kind of hibernation waiting for the next rainstorm or floodwater, prosopis trees can continue providing the framework for a woodland community.

The lack of young prosopis trees in the woodland is noticeable. Questions must be asked on why none grow. Are the woodlands dying out? Clearly young prosopis are popular with the goats and indeed their habit of uniformly eating the hanging branches to as great a height as they can reach gives a distinctive trimmed appearance to the trees.

The main strategy for survival employed by the prosopis tree is the development of an extensive tap root system. This is well developed even at an early age, and once the tree is mature the tap root may be as long as 30 metres or more. However, that still doesn't explain how the odd prosopis survives in the middle of the sand when the watertable is over 50 metres down.

Remarkably, it appears that prosopis trees, like other desert plants, have the ability to absorb the night mists directly through the stomata, or pores, in their leaves once the dew has settled. Ewan Anderson's readings showed how little dew was left in the morning on his prosopis trees, presumably because it was now all safely drawn inside. This must be useful both when the tree is young and the tap root is relatively short and when it hasn't rained for a while.

Kevin had a number of methods to help him with his research into prosopis, or *ghaf,* as the Bedu call it. Much of his work entailed physiological experiments back at Durham, where controlled work on absorption rates and so on was to be carried out over the next year. Good seeds were required and a large number were collected from as great a variety of trees as possible. The shape of the trees, the height, the girth, the number of stems and other characteristics were recorded at each site. Sometimes as many as six main stems arose from one prosopis clump, all sharing the same tap root. This cloning effect is a feature of the trees' survival. Although new genetic material is not being passed on, this vegetative propagation is important for maintaining a woodland. Kevin once borrowed a mechanical digger to see if three separate trees quite close to each other were in fact a single clone and had to dig 5 metres down before the three converged.

A common feature of many of the trees was the dead or dying stems down the length of the trunk. These trees do stabilize the sand by accumulating it at the base and thereby retarding sand flow. As the sand rises up the trunk the tree must respond by growing taller in order to maximize the exposure to sunlight of its leaves. These dying stems may therefore belong to an older canopy, which has now been replaced by a younger, higher canopy. Some of the mounds of sand under these trees are up to 4 or 5 metres deep and the trees sometimes attain a height of a further 20 metres above the sand, so they can be very large indeed.

Kevin was pleased by the great variation in the forms of the trees and that there were so many locations within the Sands to collect seeds and samples to take back to the laboratory in Durham. It is his intention to grow a series of genetically representative prosopis clones; this is considered important, since there has been no regeneration of the woodland from seed in recent times.

The process of how water is actually absorbed through the leaves is little understood and Kevin set up a series of long-term studies to determine precisely how this happens, what physiological processes occur and what effect the climate has. Many of the answers will not be immediate, but as Kevin sees this tree or its relatives as being suitable for stabilizing deserts all round the world, much has to be done before it can be 'let loose' within the scientific community.

The ecological importance of the trees is that their canopy shades and shelters the area beneath from the strong sun and the ceaseless winds to which the Sands are subjected. The woodland provides a number of temperate habitats for species that would not otherwise be able to survive in the Sands. From the trees droop three species of grey and green lichens favoured by gazelle, goats and camels. After rain, the shade allows development of a rich if patchy cover of ephemeral plants. Under the bark or beneath the dead wood lying on the ground beneath the tree, live various invertebrates such as camel neck flies, spiders, pseudoscorpions and true scorpions, ants, beetles, silverfish and ticks. The tree itself has insects dependent upon it for food as well as for shelter, such as the bruchid beetles, greenfly and gall mites that infest its leaves, fruits and flowers.

Some of the above provide food for the grey monitor lizard which burrows under the roots while two species of geckos live under the bark. The invertebrates form part of the diet of many birds which were also attracted to the fruit and seeds. These woodlands offer perches, shade, nest material and nest sites and are consequently a popular haunt for birds; during the project Michael Gallagher recorded 28 species including a bee-eater, a long-legged buzzard, a grey francolin, a spotted thick-knee, collared doves, little owls, Indian rollers, yellow-vented bulbuls, a desert lesser whitethroat, an Arabian babbler, a purple sunbird, great grey shrikes, brown-necked ravens, house sparrows and a yellow-necked sparrow and Oman's first record of the rufous bush robin. Ruppell's fox, wild cat and the white-tailed mongoose were also sighted in the woodlands and if the Wahiba bear were found anywhere it would surely be here.

These woodlands are a valuable source of shade, shelter and firewood for the Bedu, who have never over-exploited them. The balance of nature has been respected and protected in the knowledge passed from generation to generation. Wells in the eastern woodland provide excellent sources of good, salt-free water. The vegetation is valuable browse for their camels and goats and if there should be any rains the grass that grows under the prosopis trees is an added source of food (after the rain fell at Qarhat Mu'ammar the Bedu took their goats and camels to other pastures to wait until the grass had grown to its full height of 18–20 centimetres before returning. However, the onset of the pick-up truck, a direct result of increased oil-wealth within Oman, is already showing signs of upsetting this balance. A month's wood collecting can now be done in an afternoon and whole areas of the woods are cleared of naturally fallen wood and underbrush.

Although the woodlands on the eastern margin had the largest concentrations of the remarkable prosopis tree, it is also found in a belt in the central areas of the Sands and also to the west in Wadis Andam, Matam and Halfayn, which will always be a focus for Bedu and wildlife. The Government of Oman has conservation plans already, but Kevin wants to take the role of this tree a stage further. His work in Durham is showing that it has a very sturdy shoot that can cope with sand, rocks, camel hooves, vehicles and so on, although it is eaten by camels and goats. Once it has reached a certain height, however, it is protected by its thorned stems and branches. It is possible that a severe wetting is needed to launch it on its career – a major flood may have been the original trigger for the eastern prosopis trees, all of which are the same age. Kevin believes that, given the right conditions, *Prosopis cineraria,* colloquially known as *ghaf* and indigenous to the Sultanate of Oman, is a strong candidate for the afforestation of the arid and semi-arid areas which now cover one-third of the world's land surface and which are increasing by 5 million hectares each year.

Sophie Laurie is another University College campaigner who was a key member of the life science team. Her appearance belies her strength. At a meeting at the RGS shortly before we all flew out to Oman, Sophie, skinny, pale and very shy, was quietly getting on with her knitting. Elsewhere round the polished table of the Council Room, the tough task ahead was being discussed – the heat, the scorpions and the deadly carpet viper. I was greatly concerned that Sophie would not be able to cope with the rigours of the project but soon had to revise my thoughts; she was tough and fit, and able to work day and night without seeming ever to sleep. (Once we met at a Sunday lunch party on the day of the London to Brighton bike ride. 'But I thought you were cycling today to raise money for the British Heart Foundation,' I said. 'Yes,' she replied, 'I've been there and back.') In Qarhat Mu'ammar, her light was on until dawn as readings had to be done throughout the night. It just shows how wrong first impressions can be.

Sophie carried out complex work on the way plants survive in the desert. Deserts lack water and have low levels of nitrogen. This is because plant productivity is limited by the lack of water, and it is plants that are the major source of nitrogen in soils. Sophie studied the way plants overcome these two problems, especially lack of water. She had equipment sufficient to measure how quickly a plant's metabolic activity reacts to changes in the environment, such as the daily fluctuation of temperature and water availability or to that rare event, rainfall. She assessed the basic plant business of nitrogen assimilation by measuring levels of an enzyme, necessary for this process, called nitrogen reductase.

Some desert perennial plants generate enormous negative (suction) pressures to keep hold of the water in their tissues. This pressure is known as the plant's water potential. Very low pressures are vital to some species of plants because they allow retention of enough water to be active in drought conditions for long periods. These pressures are generated by the plants producing within their cells large concentrations of harmless organic chemicals which generate very high osmotic pressures within the plant, inhibiting movement of water out of the plant. Sophie measured the potentials in a number of plants using a pressure bomb into which a leaf or stem is sealed with only the cut stem protruding. The bomb generates an increasing, measurable pressure on the leaf until the pressure is great enough to drive water from the leaves out of the cut end of the stem. A small, rather ordinary looking succulent plant came top of the negative pressure league with less than −400 psi (pounds per square inch). This was *Zygophyllum qatarense* which is often green when other plants are dried up.

The water potentials varied greatly at different times of day as the plants reacted to changes in the temperature and humidity of their environment and each species measured had a different cycle. She was also able to analyse the compounds responsible for generating the high osmotic pressures. This was invaluable fieldwork allowing an understanding of how plants cope with the desert. It was not replicable in the laboratory for many reasons, one of which was that, as Sophie found, it was only possible to reproduce light levels in the laboratory at about one twentieth of those found at Field Base at midday!

Water potential may be regarded as the osmotic pressure by which a plant draws water from the soil and then retains it. These potentials can be measured in the field by using a Vinten Plant Water Status Console and data from the automatic weather station.

All animals rely on plants for food, either directly or, in the case of carnivores, indirectly. Some carnivores are unable to eat plants because they don't have the strong digestive enzymes to break down the plants' tough outer walls, so they resort to intermediate members of the food chain such as animals without backbones (invertebrates) or fellow vertebrates such as snakes, lizards, frogs, geckos and gerbils.

Studying the invertebrates were Willie and Sonya Buttiker, both experienced desert campaigners. Willie, a medical entomologist, had been studying the biology and control of insects with respect to their agricultural, medical and veterinary importance since 1948, working in Zimbabwe, Sudan, Mauritius, England, Afghanistan, Sri Lanka, Burma, Indochina, Oman, Qatar and the Philippines. He even spoke some colloquial Arabic. Such experience is invaluable on expeditions and although Willie was the elder statesman of the team he proved to be one of the most productive. His task was to survey the invertebrate populations within the Sands, working in conjunction with Michael Gallagher.

Willie pointed out that Oman and the rest of the Arabian peninsula is situated on the junction of not two but three major zoogeographical regions (Palaearctic, Afrotropical and Oriental) and an assessment of the material collected there would contribute greatly to the understanding of the geographical interrelationship between the Middle East and adjoining territories. Many invertebrates are wingless and their immobility and geographic isolation from other sand seas, particularly the Rub' Al Khali in Saudi Arabia, meant that there was likely to be a high number of invertebrate endemics to be found in the Sands.

A typical catching session lasted two hours; they took place in all parts of the Sands and each was different. On one occasion they left Taylorbase and went round the garden opposite with its palms and irrigation channels.

Most of the village around which this garden was built is now deserted, with old mud houses standing on ancient tells above the level of the main wadi that edges the Sands. Willie selected a site surrounded on two sides by palm groves and prosopis trees. Amongst the self-sown prosopis trees was an old mosque with a water channel set out for washing before prayer. The water was carried to the mosque by a low tunnel from which it emerged to run beneath a flight of long steps. The mud walls of the mosque were cracking with age and the decorated mud lattice on the arched windows was mostly broken, only fragments remaining to tell of the care that had been lavished on it long ago. Inside the creaking tilt-hinged doors, two large, broad, pointed arches supported the roof and before them lay a pile of old palm logs. In the back wall was the niche indicating the direction of prayer. Outside in the darkness, beyond the light of our lamps, a strong wind blew up from the west, causing all the palm trees to rustle and sway whilst funnelling volumes of dust down the track beside the mosque as if something powerful stirred in the darkness beyond the light.

Just within the light in the garden of the mosque circled a crowd of bats in search of the same insects as Willie and Sonya; these lived by day in the tunnel of the *falaj*, but on this warm damp night had come alive to swirl around the creaking trees. As soon as the mercury vapour catching bulb was set up, moths, wasps, tiny flies and bugs started to arrive. Grey, brown and green with an infinite variety of pattern were the moths' wings which brought them to their doom. Willie and Sonya knelt collecting busily, bottles of ether and alcohol sealing the creatures' fates. They kept a wary eye open for the lemon yellow sand scorpion, the most poisonous in Arabia, which is attracted to the light. A look beneath some scruffy bushes growing beside a wall revealed a blunt-nosed snake with orange and black markings inching slowly along the wall. This pretty creature was not spared either, nor was a pink worm snake, looking like an earthworm, which was found under some debris near the mosque's enclosing wall.

Willie and Sonya agreed afterwards it was one of the best sessions of catching since they had arrived, the result of an increase in heat and humidity and also of the well-chosen site. The catches increased as the weather became warmer with the approach of summer.

In three months Willie, Sonya and Michael Gallagher collected over 16,500 invertebrates. This is a staggering number for a desert and averages over 200 a day. They sampled at several locations in the Sands including around Taylorbase, Ras Dhabdhub (the Central Sands), Tawi Sarim (to the west), Quhayd (at the coast), and extensively at Qarhat Mu'ammar. There were always plenty of sites to investigate. Many invertebrates and small mammals used the 'plant islands' as shelter during the extremely hot daytime hours. A number of insects feed on windblown detritus and could be found in the nooks and crannies of the sand dunes. Others took refuge in the root systems of those plants that had tussocks.

The order of insects with the most representatives in the collection were the Hymenoptera (wasps, bees and ants), many of which are winged, social creatures; both characteristics may give them an advantage in the tough desert environment. The Coleoptera (beetles) were the second largest group in the collection. After each dawn their abundance was testified by the numerous trails of footprints of different sizes criss-crossing the sands; as the day progressed the windblown sand would elide these tracks, but every night a different pattern of tracks would be made. Butterflies and moths were well represented but the migratory caper white (*Anaphaeis urota*), seen in abundance on the mapping phase in 1985, was not present in 1986. Other invertebrate groups were collected in the Sands, in the wadis bordering the sands or in the wells. These included molluscs, ants, scorpions, amphibians, reptiles and freshwater fish. At the end of the project these specimens were all sent to some 80 specialists around the world for identification.

Collecting techniques varied but were undertaken day and night at all hours. The list in Willie's report includes:

Forceps – for invertebrates under stones, on the ground and in plants, etc.

Catchering – beating trees so that insects fall into a shallow tray or cloth, for invertebrates in trees and bushes.

Mercury vapour trap – nightcatching with a night light trap run off a small generator in camps, in wadis, the oases and sands.

Malaise traps – tent-like mosquito nets that funnel insects into a collecting bottle, used in camps.

Sieves – in dry sandy areas, especially in tussocks, but also in pools and running water.

Butterfly net – for insects, freshwater fish and reptiles.

Fingerclip – for fleas in rodent burrows.

Sandfly traps – caster oil coated tracing paper on plastic sheets, for sandflies and other groups of midges and small insects.

Such an arsenal was to prove effective, but conservationists might be alarmed at the quantities caught. This issue is much debated. Willie argues that these collections do not have the slightest impact on the overall populations, which are in the billions, and that small samples like this are essential for pushing forward the frontiers of science. Of the many finds by Willie and Sonya one of the most exciting was that of the dew-drinking beetles.

They found they could collect up to thirty beetles in the hour after dawn on dew-covered dunes. These beetles all belonged to one of five species each new to science. These species belong to either the genera Ammogiton or Erodius. Related species are known from the Namib Desert, Tunisia, Morocco and the north-west Sahara and Saudi Arabia, but the only previous observations of dew-drinking behaviour were ten years ago in the Namib Desert where beetles were observed to construct trenches perpendicular to fog-bearing winds. They ingested water, increasing their weight by some 13 per cent. Early morning work by Willie was rewarded by observation of beetles using mounds in addition to trenches to gather water from free dew. Subsequently, as the sun rises, the dew evaporates, the mounds and trenches disintegrate, and the beetles dig themselves in 10 to 20cm beneath the dune surface to avoid the sun and heat of the day.

Taxonomists take the view that we must know what is there, and in what numbers, to be able to monitor and predict the future welfare of the environment. Continued ignorance of the earth's living creatures would prevent long-term monitoring, and the dew-drinking beetle is as relevant to these Arabian sands as the cheetah to the African savannah. Information on the numbers and diversity of insects predict major imbalances in the kingdoms of nature, such as the swarming of locusts or the increased numbers of mosquito.

Married partnerships can be upsetting influences on expeditions, but Willie and Sonya were a strength. Everyone on the team benefited from their kindness and calmness. They too were night owls but their association with the Field Base donkeys was to try their patience initially. One night after supper we sat round a small fire in the Sands and Sonya told us the story.

'We love donkeys and were pleased to find five healthy-looking specimens wandering between the tents and the prosopis wood at Qarhat Mu'ammar. During daytime they were just lovely to watch, standing in a tight circle dozing in the shade of the trees. It was during the night that they seemed to change character. We had barely settled into our camp beds when the donkeys began to come to life. First they exchanged their long, weird brays of greeting from hill to hill and from wadi to wadi, then hunger and thirst made them search for food and water; their preferred hunting ground was around our tents. Nothing left outside was safe – washtowels, buckets and every piece of plastic or paper attracted their curiosity. Also, they preferred the tent walls to any prosopis trunk to scratch their tick-infested hides and a few times we expected the tent to collapse on top of us. Every night the donkeys' chorus and ballet got us out of bed. We tried to chase them into other areas, calling, begging, threatening, but after a discreet absence of a quarter of an hour they were back again.

'It was during the penultimate night of our stay at Qarhat Mu'ammar that William got quite wild and jumped out of bed barely mastering his temper. I listened wide awake and somewhat anxiously when suddenly, instead of a thundering voice, there was a gentle coaxing one, begging the donkeys not to be frightened. The animals didn't trust this extraordinary change of mood and vanished in the woods, not to return that night, but William was bursting with excitement. In the light of the moon he at last had discovered eye-frequenting moths sitting in a neat circle around the eyes of our tormentors. The only other record of these in Oman had been from Michael Gallagher, who had the opportunity to observe several feeding eye-moths in the vicinity of Salalah, where two specimens were captured. For the Sands it was a novelty.

'William had seen these during a WHO research programme nearly 30 years ago. He was then studying malaria-transmitting mosquitoes in Cambodia. To find the nocturnal anopheles mosquitoes in the jungle villages, tame water buffaloes and domestic oxen were investigated. Surprised scientists noted that not only mosquitoes visited the hosts for feeding, but moths of various sizes came flying out of the thickest jungle and settled on the edges of the eyes, sucking up the 'lacrimal' fluid and sometimes also blood from the open wounds of the placid bovines. After this extraordinary discovery, more observations and investigations were carried out systematically over the following years.

'So when darkness fell next night and our last chance of collecting arrived we did not want to risk any disappointment. With the kind co-operation of the soldiers, we tied two donkeys with a thick rope to a tree near the kitchen area. Naturally the prisoners got all our attention as well as a bucket of water and some biscuits. We all sat in a circle on our heels round the fire, waiting for the moths to appear. Suddenly they came flying by the dozen out of the *ghaf* woods, to the pure delight of the entomologists as well as the spectators. As we had seen them in Asia, the moths began to seek out the eyes of our donkeys and went on feeding until shortly before midnight. This gave us ample time for a good collection of the strange eye-moths.'

As the fire died away into embers, the significance of the find was explained. These ophthalmotropic moths, as they are called, are eye parasites of wild and domestic cattle and horses. In West Africa they have proved to be the carriers of diseases such as keratoconjunctivitis epidemica and possibly others such as rinderpest.

Willie was to find a number of other insects that had medical significance. No bilharzia snails were found in the sands but they were present in the flood water of Wadi Beni Khalid. The 'desert fly', *Musca sorbens,* is a vector for trachoma and it was found in all corners of the sands. Twenty specimens of the sandfly *Phlebotomus papatasi* were discovered in the sands – a transmitter for leishmaniases (a rare skin disease, unlikely in the Sands as there was no second host). The black fly *Simulium* spp., found in the Wadi Batha waters, is a vector of ochocerciasis, the river blindness found in Africa and south-west Arabia.

From a veterinary standpoint, Willie and Sonya found a few invertebrates worth considering. A number of scorpions – including *Leiurus quinquestriatus*, the most poisonous – was found. Tick-borne diseases are a potential problem for both man and livestock and a large number of ticks was collected and checked for viral disease agents. The midges *Culicoides,* which have been known to be a vector for bluetongue in sheep, horse sickness and cow fevers, were found. The parasitic mite which causes mange was seen in camels, goats and sheep, and blood-sucking lice, which cause distress to their hosts, were quite widespread.

In their hearts Willie and Sonya are conservationists and they realize that the vast amount of invertebrates are but indicators of the state of health of any given environment. Future ecologists will be calling for an assessment of those invertebrates most sensitive to change and to varying ecological zones, or biotopes as they are called. Willie and Sonya believe that the collections made by the team are but a very small percentage of what will eventually be found. The isolation of the Sands will show many of these invertebrates to be endemic to Oman or even just to the Wahiba Sands. The wadis, the hills and the sands harbour a high percentage of unique species of small animals and, to quote Willie, 'those could be threatened by indiscriminate disturbance or even destruction of their fragile biotopes'.

Seven areas of the Sands were identified as being biotopes of particular importance to the entomologists: the *habl* dunes in the high Wahiba, the network and barchan dunes in the low Wahiba, the central prosopis woodlands, the eastern prosopis woodlands near Qarhat Mu'ammar, Wadi Batha near Bilad Bani Bu Hasan, Wadi Andam near Mudhaybi, the aeolianite formations on the sea coast and the escarpment at the western fringe of the Sands. All of these areas contained a number of characteristic invertebrates, many of them new to science.

Michael Gallagher was another researcher with nocturnal habits. While some slept, the distant glow of a hand-held kerosene pressure lamp marked his progress during the midnight hours, crossing dunes or wending slowly through the woodlands, pausing frequently as he examined every bush, burrow or tree-trunk. Among the ridge-dunes above Mobile Base 1 near Ras Dhabdhub on a cool, moonlit night in January, he paused to examine a nocturnal gecko of the genus *Stenodactylus*. These are 'still-hunters', which stand frozen on upright legs for minutes on end before running to

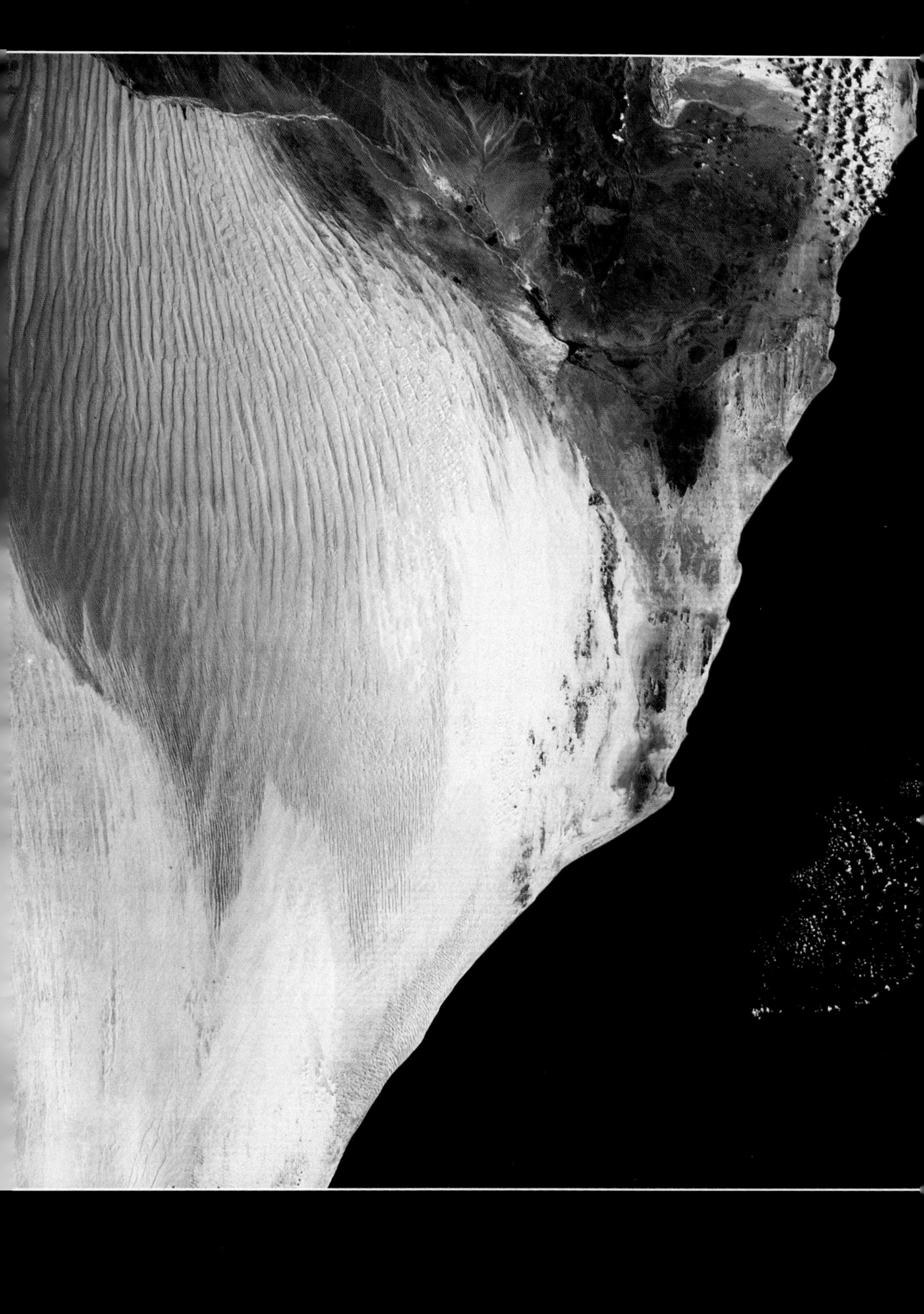

OFFICIAL CARRIER
OMAN WAHIBA SANDS PROJECT 1985 86
BP

another position to stand, watch and wait – and pounce upon any unwary moth or ant. As he knelt to peer at it more closely, he noticed out of the corner of his eye, at the very edge of the lamplight, the zigzag form of a viper hiding under a bush. While the lizard scampered off, unaware of its narrow escape, Michael photographed the snake, then lassoed it with his snake stick and put it in his bag.

Next morning an assorted party from Taylorbase arrived and grouped round, cameras ready, as Michael released the viper on to an open patch of dune sand. Never had most of us moved so fast. Having been told that it was a sidewinder, most of us expected a leisurely demonstration of this special method of locomotion on sand. Instead, with its head held high and with only two parts of its belly in contact with the hot ground at any one time, the horned viper *Cerastes cerastes* almost flew across the dune, with spectators running to keep pace.

This viper measured 58 centimetres and was prettily patterned. Its 'horns' of scales were indistinct above the eye, but as all vipers have a pair of hinged, hollow, needle-like fangs at the front of the upper jaw capable of injecting large quantities of fatal venom, this one was sufficient warning to everyone answering the call of nature among the dunes and bushes at night to examine their surroundings carefully. However, Michael pointed out that fatalities due to snake bite in Oman were rare and that vipers didn't attack unless sorely provoked – which was comforting for those who slept on the sand at night.

Michael was to increase his records of reptile and amphibian species from 10 to some 20. These included the Dhofar toad, the oriental toad, several species of geckoes including the common *Pristurus minimus,* several lizards including the toad-headed agamid which liked to bury itself under the sands, *Uromastyx microlepis,* the spiny-tailed dhab, the Haas' spiny-footed lizard, the skink (*Scincus mitranus*), the grey monitor lizard, the rare 'worm lizards', a thread snake, Jayakars's sand boa, the common tree snake, the horned viper and the carpet viper. Not bad for a patch of sand.

Amphibians don't like cold weather. At the start of the project the days were cool and dry and there were clear skies by day and night and a bright moon. This tended to make the shyer nocturnal lizards keep cover. Dust-raising winds sometimes persisted after dark, but this did not seem to deter the larger geckoes like *Stenodactylus. Acanthodactylus,* the spiny-footed lizard, used to wait until after 9.30 p.m. before venturing out. During February, warmer temperatures and some rain were accompanied by more reptile activity. By March, when the ambient and ground temperatures were higher, the monitor lizards and the *Uromastyx* became more active and the Dhofar and oriental toads were both singing in the irrigated gardens of the villages.

The diversity of habitats within the Sands, including the gravel wadi floors, the sand dunes, the interdune hollows, the aeolianite, the *ghaf* woodlands and the flats and hills of the coast, will always ensure a good representation of reptiles and amphibians; much more sampling work needs to be done and interested parties should make a beeline for the Natural History Museum in Muscat.

During the project there were three species of lizard that were of particular interest. One was *Uromastyx microlepsis,* the spiny-tailed dhab which is found over much of Arabia. These are striking creatures, being large with a tail armoured with lumpy spines. They live in burrows in firm sand and silts and the spiny tail is said to help them anchor themselves in their burrows when attempts are made to pull them out. There were plenty of likely burrows in different parts of the Sands but very few dhab were seen and Michael was anxious to establish that they were still extant in the Sands. Returning one day to Mobile Base 4 near Shaqq after another unsuccessful search for these animals, he caught sight of the characteristic barbed tail in a bush near the kitchen; further inspection revealed first clawed feet, then the detached head, but of the body nothing remained for it had been eaten by the soldiers as an *hors d'oeuvre.* This substantial lizard is favoured by some Bedu tribesmen as a desert food.

Paul Munton found one crouched in his path in the eastern Sands and was able to pick it up for it did not move as he approached, being cold and moribund. He kept it in his Land Rover for three days but it only really woke up when the sun brought the temperature in the cab up to over 39°C. At these times it would lose its dull flattened look with its belly touching the ground and instead would support all its weight upon its substantial legs with an alert gleam in its eye. At such times it would also show fear when Paul moved suddenly. It was clearly too cold for the animal so it was returned when warmed up to a suitable habitat to find a burrow and await the coming of the heat of summer.

In addition there were two discoveries to go into the record books. The first, the nocturnal gecko *Stenodactylus khobarensis,* had never been found in Oman before nor so far south. It had been named in 1957 from a specimen found at Al Khobar, in eastern Saudi Arabia, and had been found subsequently only in Bahrain and the United Arab Emirates. Measuring only 5 centimetres from snout to tail-tip and almost indistinguishable in the field from others of the same genus, this species has a most unusual liking for *sabkhas* (salt flats) and beaches, so during Michael's many explorations near Qarhat Mu'ammar he spent long night hours searching the dry and apparently inhospitable *sabkha* nearby and found a single specimen towards the end of February. This raises the intriguing possibility that it occurs on *sabkhas* throughout the country and even on the great but evil Umm as Samim quicksands at the edge of the Empty Quarter.

The second discovery was *Diplometopon zarudnyi* a 'worm lizard' belonging to the group Amphisbaenidae. This intriguing quirk of natural history is neither a lizard nor a snake but is related to both. Although it is mostly subterranean in its habits, Michael noticed its distinctive trails on the sand surface early on in the project. The problem was to see one on the surface. The first specimen from the Wahiba was attracted to Willie and Sonya's light trap in the deserted village opposite Taylorbase, but during the night towards the end of March Michael was lucky enough to be able to watch one searching for ants near Tawi Sarim in Umm Qishrib, part of the central prosopis woodland. His subsequent photographs show the remarkable way these creatures move on the surface, curling the pointed tail to one side, pressing the tip into the ground to assist the lunge forward, then repeating the action by curling the tail to the other side. To go underground, *Diplometopon zarudnyi* will press its shovel-shaped snout into the sand and with a quick, worm-like motion will soon be deep in the sand.

One rung up on the food chain are the small mammals that scurry around looking for scraps of food, detritus, dew-drinking beetles or even a tasty amphibishaenian. Luckily there seemed to be plenty of mammals about, particularly in the vegetation mounds and woodlands. Michael Gallagher was to co-ordinate this part of the survey, organizing lines of traps, which needed checking every three hours, at all the bases as well as keeping an eye on all the other groups including the birds.

The three years of drought had no doubt reduced the number of bats, rats and gerbils. Cheesman's gerbil was widespread throughout the Sands, albeit in small numbers. This gerbil, related to those you might find in the petshop, has evolved a useful mechanism to cope with the lack of water. When we urinate we dispose of both excess salts and water, but the kidneys of this gerbil recycle its water so that it is all reabsorbed into the body, leaving only a very strong salty substance to be urinated. In this way the gerbil can survive by drinking the dew that settles on the plants.

Michael was able to collect three types of bat: the trident bat, the Persian leaf-nosed bat and the common pipistrelle, which often skimmed over one's head around the mess tent at Qarhat Mu'ammar or at one of the light traps. The black rat and the house mouse were collected at Taylorbase – just two members of the community that beat a path there for food or soft packing material for the nest. The dripping taps of the water tanks were also popular by the look of the myriad little footprints. Although the burrows of Sundevall's Jird, *Meriones crassus,* were sighted in the sand hummocks of Wadi Mattam, only one came to Michael's traps. Jirds are robust, rat-like gerbils and this species typically lives in colonies in complex tunnel systems beneath bushes. But the sands had a great variety of footprints indicating nocturnal activities by others and suggesting the small mammal population may be larger than our traps gave us to believe. Said Jabber showed me footprints of a hedgehog and small mice which Mike thought might be spiny mice or jerboas.

Michael used a variety of methods to catch his specimens. These included rat-sized Sherman traps which were light aluminium boxes, triggered when the prey nosed a little lever attached to a chewy bit of oatmeal, peanut butter, fish or left-over chapati. Standard mouse traps were also used and mist-nets made of very fine but tough thread were put out at night to sample the bat populations. There were others apart from Michael interested in the contents of the mist-nets at night; a large gaping hole in one indicated that some carnivore had found his hunting made easy. This was probably a fox, the main focus of interest of Ian Linn.

Ian, a zoologist of renown from Exeter University, has a long track record studying carnivores which goes back to 1962, when he published his discoveries about the weasel. With a huge white mane of hair and beard and bright eyes that sparkled when telling jokes, Ian was a popular character about the camp – although not at supper, for he had a good appetite; with the communal trays it was sometimes a question of the survival of the fittest.

At Taylorbase Ian commandeered the services of the whole team to help him collect information about the larger mammals. These are generally easier to identify and all members were issued with a list of them to look out for. This list was formidable and very exciting indeed; it seemed incredible that such a desert could support carnivores so high up the food chain. It included mammals such as *Canis lupus,* the wolf; *Ichneumia albicauda,* the white-tailed mongoose; *Vulpes ruppeli,* the Ruppell's sand fox and *Felis silvestris,* the wild cat. The instructions were quite clear: 'Leopard – big spotty pussy, smaller and paler than the ones in the zoos. Big rounded feet, so leaves rounded track. Arabic name *nimr*. Very unlikely'.

When half way through the project we were visited by our Patron, HRH Prince Michael of Kent; we took him on a trip around the Sands, including the stretch of coastal dunes running south from the fishing village of Khuyamah. About 30 kilometres from this village is a rock, which because of its shape was given the name Monument Rock. The party had just reached this spot. Prince Michael and Mike Holman were in the lead Land Rover driving up and over the rolling dunes when they suddenly spotted a dog-like animal running behind a dune. They gave chase, leaving the rest of the convoy way behind, and followed the animal for several kilometres. They were able to get near enough to see that it was a wolf, but just as they were closing in on it they drove into a deep hole of soft sand and became well and truly bogged down. The wolf disappeared over the crest of a nearby dune, never to be seen again.

The two Michaels rejoined the convoy, after a bout of heavy digging, triumphant at being able to add to the sightings on the list.

During the project we ticked off the greater proportion of Ian's list. The variety of habitats provided by the Sands, particularly the prosopis woods, were attractive to a variety of mammals and the fish left by the seashore communities provided a great deal of food for a community of red foxes which we often saw when driving down the coast to Quhayd.

Of all our sightings the most unexpected was the first of two white-tailed mongooses, which so far was only known to live in thick bush near Dhofar and the Batinah coast. Ian, while trapping for sand cats, had somehow managed to entice the mongoose into the baited trap that had been put among the brushwood of the prosopis woodland. As it was the first of its kind seen in the Sharqiya, Ian was interested in its behaviour, its territory and where it lived.

A few days later Ian was delighted to catch a second mongoose. One night he took a small group round his territory in the woods at Qarhat Mu'ammar, radio tracking for the animals he had collared. Each had been fitted with a small battery transmitter and a short flexible aerial, neatly attached by a dog collar. The radio signals these emitted could be picked up by a radio receiver which Ian carried around with him, together with a hand-held aerial. It was remarkably accurate.

We left at about 10.00 p.m. in one of the open-top vehicles, Ian looking like a crusader. His huge mane was now tied up in a *masarra* and his beard was flowing. He took the wheel and sped off into the night, with us hanging on with very white knuckles, but Ian knew his ground and took the vehicle round a set course that was by now well worn. He was not in a happy frame of mind that night; when he began his studies this area had had a thick undergrowth of scrub and dying material, just the kind of habitat night creatures like. Ian had been back to the main base camp for four days to rest and to attend workshops where the scientists discussed the progress of their work. On his return he had found that the woodlands had been cleared of all the undergrowth in large quantities, just as if they had been swept by an enormous broom; a number of Bedu had come with their trucks and stocked up on wood-fuel. This disturbed Ian as in the past the Bedu wouldn't have taken so much at one time. Now, the animals he was studying had lost their habitat in one fell swoop.

We pulled up at a narrow gap in the woods and turned into it. There were just a few centimetres to spare on either side of the vehicle as we nosed through into a small clearing. The lights picked up a little label on a tree and Ian made a note of the time. Telling us to be very quiet, Ian took his aerial and bleep box from the back of the Land Rover and, carrying our torches, we followed him into a small clearing in the woods.

Ian put his earphones on and started waving the aerial above his head.

His radio receiver made a hissing sound as we waited for it to pick up any signals from the transmitters on the two mongooses Ian had caught. Suddenly the box bleeped – quietly at first, but regularly. Ian shifted the direction of his aerial and the bleep became louder. By moving the aerial round, he was able to determine the direction of the animal relatively accurately and now noted the bearing – the mongoose was about 20 metres away and moving over to a hill.

The patch of woodland that Ian had marked out was about a kilometre square and by driving round it and taking readings at various intervals he could build up a picture of where the animal lived, how far it travelled, whether it left the territory and what it might be feeding on. By tracking two animals he could show how their territories overlapped – he had successfully done this earlier in the project by studying the foraging habits of two male Ruppell's foxes at Ras Dhabdhub, both living within a small area and, in this case, the way they competed for the remains of chapatis and curry in the camp rubbish tip.

We were lucky enough to catch a glimpse of the mongoose as it darted behind a small sandhill making for its burrow. As we continued our search for the other mongoose while waiting to see if the first one reappeared, I began to have the greatest respect for this zoologist who was dedicated to learning more about these shy creatures; his commitment to spending most of the night on his own watching, waiting and listening was formidable. His estimation that the woodlands sustain a population of white-tailed mongooses had direct implications. To have mammals like foxes, sand cats and mongooses at the head of the ecological chain, who in turn are only threatened by man, is indicative of the rich resources of the Sands and their margins. The plant and lower animal communities can sustain healthy predator populations, particularly in the prosopis woodlands.

We were able to get a glimpse of the life of the predatory mammals that night and, when added to that of the invertebrates, amphibians, reptiles, small mammals and bats, we realized that the woods were a seething mass of activity and that there were probably many eyes watching us as we sat on a small hill waiting for the bleeps. Something that Said Jabber had told me became clear to me; that to see life in the Sands, it is not a question of just looking but also of understanding. By merely looking you don't necessarily see anything. We were lucky that night to be able to benefit from Ian's 30 years of accumulated knowledge.

I was always impressed by the Bedu's keen eyesight. When Said took us across the Sands he was able to attract our attention to all manner of details, not least the birdlife which numbered some 97 species. Birds eat plants, invertebrates and some vertebrates depending on dietary preferences and the plentiful supply of invertebrates was responsible for a large part of the avian population. Michael Gallagher is a renowned expert on the birds of Oman and has written a popular and authoritative book which is published both in Arabic and English. Having Michael on the team was like having a 24-hour, easy to use, well-referenced, well-indexed, answer-back living encyclopaedia. There is no other ornithologist in the world with Michael's experience of Oman and we were very lucky indeed that the Minister of Natural Heritage and Culture, Sayyid Faisal bin Ali Al-Said, had such an interest in the wildlife of Oman and so allowed Michael many days away from his desk at the Natural History Museum.

Michael was able to divide the sightings. Birds are either breeding or non-breeding and are resident or just passing through (migrants):

Non-breeding winter visitors	27
Resident breeders	33
Migrant breeders	8
Passage migrants	29
Non-breeding summer visitors	0
	97

But only a very small number were seen in the Sands and half that number were true sand birds. Of the observed habitats numbers were as follows:

Sand	9
Desert trees	20
Rock outcrops	3
Wet wadis	9
Cultivations and plains	23
Sea and shore	33
	97

Michael was delighted to learn of a few new species. The little green heron was seen on some rocks in Wadi Halfayn, near Hajj. The lappet-faced vulture, a huge scavenger, was seen flying over Taylorbase and probably had a nest somewhere in the Sharqiya. The long-legged buzzard, one of the birds we saw in the Sands, probably breeds in the prosopis forests. The spotted thick-knee was heard calling at Qarhat Mu'ammar in February and Ian saw a pair in the evening at twilight; this could be the most northerly record for a nesting pair. A huge roost flock of some 130 great black-headed gulls was spotted on the south-east coast in February – the largest recorded in Oman. The beautiful collared dove found calling in the woodlands may breed there while the booming call of the Bruce's Scops owl, again in the woodlands, is a most distinctive sound which for a moment raises the hairs on the back of the neck. The house sparrow and the yellow-throated sparrow were also found in the Qarhat Mu'ammar woodlands, their most southerly location known so far. The finger of *ghaf* extending to the south and the rich resources of the coast provide an ideal habitat and Michael believes there will be new sightings along that coast road for quite some time to come.

Within the Sands themselves the four resident species are the coronetted sandgrouse, the delightful hoopoe lark, the black-crowned finch lark and the brown-necked raven; most of the others that were seen in the desert were likely to be migrants. The *ghaf* woodlands offer perches, food, cover, nest material and nest sites and so will always be popular while in the Barr al Hikman, just to the south of the Sands proper, the mudflats are extensive and offer one of the best feeding grounds in eastern Arabia. This area deserves a special expedition of its own and I know Michael has had one planned for some time now. The south-west monsoon winds blow from about mid-May to mid-September and the currents induced by this wind cause upwelling of cold water from lower levels of the Arabian Sea. This water is rich in nutrients and is one reason why marine life in the coastal seas of Oman is so abundant. This richness also benefits the mudflats of the Barr al Hikman. Flocks of greater flamingo (*Phoenicopterus ruber*) and thousands of waders use this area on their way north to Eurasia, and clearly this wetland is of international importance; much more work needs to be done to study its biomass – that is, the number of living organisms to be found there – and those that exploit it. Furthermore, the Sands differ from other parts of Oman because they lie on the migratory routes and birds follow the wadis and coastline. Consequently there is much to be done by any future ornithologists visiting the Sands or their margins, not just to add to the list of sightings but also to contribute to the body of knowledge kept by the Natural History Museum in Muscat on breeding patterns, behaviour and the peaks of migration.

Watching gazelle was one of the great pleasures of the project. Sometimes one would come upon them close to the track road, emerging out of the morning mist, startled, walking stiff with apprehension. In some parts of the Sands they would be seen fleeing from vehicles, their tiny bodies rising and falling, made visible only by the pale rump patch beneath the raised tail. The flight distance from motor-car noise was more than a kilometre. To watch the gazelle it was necessary to find a high point in the southern dunes and scan with binoculars the country round about for several minutes before a movement would give one animal away, whereupon an intensive look in its immediate vicinity would usually reveal perhaps one or two others or sometimes ten animals or more. Then the observer could watch for hours and obtain some idea of the quiet lives of these creatures as they browse silently and rest in spots sheltered from the wind. Only in the presence of man do they have to run and be afraid.

Ever since Paul Munton spotted those gazelle from the Defender during the mapping visit in January 1985, the numbers of gazelle in the Sands had been a puzzle. There is no doubt the Sands form a refuge, particularly so during the days when they were hunted in large numbers. On the plains they never had a chance against the fast four-wheel-drive vehicles and automatic rifles but in the Sands the odds were tipped more in their favour. Hunting is now illegal, but Said once showed me tracks of hunting parties as we crossed near Ras Dhabdhub and he confirmed that a small amount of hunting still goes on. Wilfred Thesiger mentions seeing 'many gazelle all very wild' when he crossed the Sands from Wadi Andam via Tawi Harian and it was in that same area that Paul spotted a few more during our flight.

Paul's estimate from that first visit suggested some 2,000 gazelle but a problem is to distinguish between the Arabian gazelle (*Gazella gazella arabica*) and the sand gazelle (*Gazella subgutturosa marica*), known locally as *rheim*. Said had confirmed that we had seen three *rheim* during the mapping phase. All team members were asked to be particularly diligent when taking gazelle records; gazelle are variable in colour and markings and this is often made more difficult by the varying intensities of lighting conditions. Some gazelle have no side stripes, while others have very thick black ones. The diagnostic features for the Arabian gazelle are the spiral (lyrate) horns that point forward at the tips and the black spot just above the nose or a totally dark face. The rarer sand gazelle has lyrate horns which point forward in their entirety and a white face with no visible black marking. In the haze of the Sands it is easy to confuse the two.

These animals are important in the same way the tiger is important in Nepal – as an indicator species that confirms the conservation status of the land. Aerial flights were undertaken in January and March of 1986 to survey and count livestock and gazelle. Because gazelle were so small and blended so well with the Sands, it was found that they were only visible early in the morning and late in the day when their long shadows made them easier to spot. For most of the day they were so difficult to see that systematic survey data was of no use for calculating the total population. Although observations by project members on the ground showed that gazelle occupied the sands as far north as the 30 kilometre mark on the M2, it was clear from aerial survey and ground observations that there were concentrations of the animals in the centre of the sands where there were few Bedu houses, and on the borders of very soft sand south of Ras Dhabdhub where vehicles could not be driven. The only way to determine the true numbers and localities of the gazelle will be by a systematic survey throughout the Sands on camels and with assistance from the Bedu at the same time as an aerial survey with a little help from pilots at Seeb. That would be an exciting expedition for any student studying natural sciences at the Sultan Qaboos University. I know Paul would be the first to put his name down.

Towards the end of the project I was sitting with Paul around a fire one evening, sharing experiences and his current vision for the Sands. As Director of the Biological Programme he travelled extensively through the Sands in the course of vegetation surveys, and whilst visiting his dispersed team his only contact with the outside world were three radio calls a day to base and perhaps the odd Bedu. He would frequently wake alone amongst the dunes to the golden light on the eastern horizon, the first of the new day; he would emerge from bedding, shiver on placing a bare foot on the dew-wetted sand and listen to the silence, remembering that in the remotest villages sleep would have been interrupted by the call to prayer. The tropical dawn, which brings the light so much faster than in temperate latitudes, revealed the beauty of the world.

Paul's sleeping places were small flat areas of firm sand amongst the smooth shapes and contours of the dunes, a repetitive smoothness extending as far as the eye could see in all directions, broken only by the few scattered shrubs on the dune tops. The smoothness was violated by the impression of the tiny feet of beetles, hunting spiders and scorpions, busy even in the cool of the night. A few metres away that smoothness had been excavated the previous day by a smooth scaled skink, of the sort with a body marked on both of its shiny sides with five bold blood-red patches like human fingerprints. This warmth-loving creature was not active at dawn but would emerge towards midday when sand and sun would warm its body and renew its energy. Then, too, the sand would be brushed into shallow whorls by the feet of combfooted lizards as they sought out their diet of unwary insects. Within 100 metres or so of the sleeping place would be found the heart-shaped footprint of the Arabian gazelle, a single male, or perhaps a pair of females, that had passed by quietly in the night. Each of the two pointed digits of each foot contributed to one half of the heart. The tracks followed the dune contour, deviating where the animal had been attracted by a bush and in some places encircled it, showing where the creature had spent some minutes eating the shoot tips.

Every day the dune shapes change a little or a lot depending on the eye of the wind and its liveliness. Sometimes a mist of sand blows along the ground no higher than the ankles, like a fine tissue mimicking the contours of the sand surface, at other times the sand blows level around the thighs. When the wind is severe the sky fills with sand grains and darkens to grey; the sun becomes a pale disc and sand streams from the top of dunes in a dense fountain. It may blow all night. The next morning the smoothness and the rounded contours are perfect, the prints of all living things elided. The dunes are like the rounded parts of the human body, shoulders, thighs, the ridge of the collar bone, buttocks, curve of the back, the shallow parabola of the stomach. All these parts, welded together in fleshy tones of tan, apricot or white contrasting with the blue of the early morning sky; mile after mile of such coloured forms stretching into the distance, form an infinite image that impressed all who camped in the Sands.

In the Sands there is a view from horizon to horizon and one of the pleasures of the day was to watch the cloud patterns change: clouds like disembodied feathery wings; low puffy cloud materializing in the heat of the day, hinting of future rain but dissolving before dusk. Sheets of cloud from the north promising rain but revealing themselves as worn-out fronts, dry from dispensing their water too liberally over the Mediterranean and northern Arabia. After three years of such days without rain, in the third week of February 1986 there came a wind from the south, tiresome, blowing the sand and dimming the sun to a flat disc for most of the day; at night lightning flashed in the north and west, and before dawn the promise was fulfilled. Then was the time to enjoy the triennial luxury of opening your mouth to drink the rain, joining with the earth around you, with the trees and every living thing.

The Koran, in the chapter entitled 'The Cow', promises man death, famine and loss but elsewhere speaks of God's generosity, for it is He who sends down blessed water which brings forth trees and their fruits and grass giving life to the dead land. He has covered the earth with desirable plants. It also says (written in AD 710) that men rejoice when it rains and so it was in 1986. When the wadi flow reached the village, the water creeping in brown, froth-edged fingers across the grey gravels of the wadi bed, prelude to a broad torrent, families sat together singing on the banks whilst the moonlight was interrupted by scudding clouds. Where cars and trucks were stopped by the wadis men joined hands by the water's edge and sang and danced together until the water had subsided enough for them to cross over.

The dunes darkened with the rain, their softness gone. In places the sand broke away in slabs and slid a little way down the dune face as if wishing to be for one day like rocks in the mountains which, as the Koran says, fall down for fear of Allah. The Bedu dig down, measuring with a hand's length, a forearm's length or a whole arm the generosity of the Almighty, predicting the growth of grass and plants to be expected, when it will come and how long it will be available for their livestock, when they should move, how far, and for how long they should stay, when they should herd the goats and sheep or load their trucks.

So the Bedu leave the woodlands to allow the trees a rest from three years of cutting and the battering of their long poles which makes leaves and fruit fall for the goats and sheep. Whilst they are away the grass will grow under the trees, safe from the trampling of the pointed hooves of goats and sheep. When they return there will be standing hay for the livestock and ripe seed ready for the next rainfall.

The biologists worked to understand the changes in the vegetation and the effects upon the creatures of the Sands. When pastoralism was being considered, work on it was undertaken jointly with the Arabists on the project to chart and assess the significance of the way the people of the Sands lived, using the available resources in times of drought and the changes that took place in the time of plenty after rain.

CHAPTER SIX

The Sand Guardians

... the men of the land, rather than the land of the men, were my main object of research and principal study.

William Gifford Palgrave 1826–88

Dhiyab's eyes are as keen as those of his namesake the wolf. The sun was low over the dunes when one day he saw a very large bird land on the ground not too far away from his Land Rover as he drove through one of the main swales. Mike Gallagher had asked that project members spend a little time to take a good look at large predators and make a description of them. Keen to contribute to Mike's inventory, he stopped the vehicle and turned off the engine.

Dhiyab slowly opened the door and got out to approach the bird on foot. The sand was still burning hot but, even though he was barefoot, he didn't notice. Holding his breath, he snatched a photograph but he knew the bird was too far away for it to be of any value. As he drew closer he could just identify the pale underwing and distinctive shape of a long-legged buzzard (*Butes rufinus*). The bird took flight, albeit reluctantly, soared off in a wide arc and landed in the shadows on the west side of the valley. Dhiyab followed, camera in hand. The buzzard remained just out of reach but was clearly reluctant to move too far away. As dusk fell Dhiyab replaced the lens cap on his camera and returned to the Land Rover. It was now time to find a camp for the night. He was alone and he was keen to make contact with the local Bedu. Also his tummy was beginning to rumble.

Just as he was climbing back into the cab, Dhiyab noticed a small form in the sand 5 metres away. Grey-brown and isabelline fur, black and white tail, parchment ears and stick legs – a hare, usually protected by its camouflage and swift flight but in this case no match for a desert bird of prey. The eye sockets were bloody. The nearby sand, as ever, told the story. Here was evidence of the fatal events. The footprints of a running hare turned to a confused scrabble where it had been struck down by the feathered thunderbolt. There were deep imprints of wingtips where the victor had crouched over its prey tearing at the eyes. Spattered blood lined a trail where the body had been dragged along, perhaps to seek cover from some nearby *Calligonum* bush as the sound of the vehicle interrupted the feast. Too late for the hare and now too late for the buzzard.

Dhiyab picked up the body with the hand of a scientist and started to think about preserving it for Michael's records. Hunger displaced recording priorities however, and turned the fate of the hare from pickle to roast. He had been travelling for four days, visiting some of the Bedu households he passed, and had stayed the night at two houses when his hosts had been especially pressing with their hospitality. His food there had consisted mostly of dates, camel milk, rice with onions and some dried fish, and of course endless cups of coffee which oiled the lively exchange of news. On the other nights he had slept on the bare sand beneath the stars. He was now looking forward to a substantial meal, a friendly chat and more coffee – but suddenly there were complications. If he ate alone the hare would be a good substitute for tinned sardines; if he could find a Bedu encampment for the night then the hare would be a welcome contribution to the evening meal and compensate for the inconvenience of entertaining a stranger. Would the Bedu accept meat killed by a bird without the proper ritual slaughter? He knew the proper procedure would be to cut the hare's throat and pronounce the necessary prayer even though it was already dead, but the night was closing in and it would take time to find his sharp knife. He still had to gather firewood in this treeless swale if he couldn't find any Bedu.

Dhiyab compromised; cutting its throat with a knife he said the *Bismallah* aloud, grabbed the lukewarm hare by its thin legs, dropped it into the back of the Land Rover and set off southwards to where the Bedu had recently moved due to the recent rains. The track was well marked and he paused only to collect some dead sticks of *Calligonum*. He drove for half an hour before stopping on a dune crest and seeing a light in the distance. This was not the dipping yellow glare of headlamps, but the welcome glow of a fire. Dhiyab turned the vehicle to cut across the low dunes with clumps of grasses and bushes newly greened by the rains. This was an obvious site for the Bedu to set up a camp. A *Calligonum* bush provided shelter for two figures who rose to their feet as he halted and climbed down from the intrusive vehicle. He felt he was back in the fold and the potential loneliness of the Sands evaporated immediately.

Resting camels grumbled as he passed between them. Dhiyab did not know these two Bedu, although they had heard of him. From a distance they exchanged the ritual greetings of the desert; now a formality, such greetings in the past had been important passwords to signify friend not foe. A tiny coffee pot was produced, ancient and soot-blackened, with a broken lid and a handle repaired by wire. This tool of hospitality had seen active service. More wood was thrown on the embers. The firelight shone on the face of a middle-aged man with a long grey beard, framed in layers of shawls against the desert cold. Close by a boy crouched over the fire warming his hands, his eyes streaming from cold and smoke, which was now billowing out from the damp wood. Dhiyab laid the hare down by the fire and they all huddled round.

They explained that they had arrived from the south that day with their herd of 25 camels. The women and children would join them in a day or two, travelling by pick-up and bringing the goats and tent materials. For the time being they had only the bush and a small canvas windbreak for a shelter. Around them lay their saddles and a few blankets. Beside the fire were two bowls of freshly drawn camel's milk which was to be supper, along with some dried dates and perhaps some rice.

Dhiyab recounted the story of the buzzard and the hare and assured them both it had been freshly killed, but admitted that he was unsure if it was strictly *halal* (lawful) or *haram* (forbidden). They hadn't eaten meat for weeks and looked anxiously at the corpse.

'Was it already dead,' asked the older man.

'I think so,' replied Dhiyab guiltily.

'Did you cut its throat?' they asked.

'Yes.'

'But you did say *Bismallah*?'

'Certainly!'

'Then it is *halal*,' they both exclaimed. Dhiyab was most relieved.

By ten o'clock, with the moon overhead, the three were crouching down to a delicious meal of hare stew, rice and camel's milk. They ate in a silence interrupted only by appropriate noises of satisfaction. The hare had been good value. Long after Dhiyab had finished, the old man was still cracking bones to scrape out the last traces of marrow. Dhiyab scraped a hollow in the sands and pulled his blanket over him. He was at home with the true guardians of the Sands; Taylorbase and Mintirib were a long way away.

Drawing together the Bedu life in the Sands and the new life of the expanding villages was one of Dhiyab's reasons for being there. Dhiyab, otherwise known as Dr Roger Webster, is an anthropologist from Exeter

University and an honorary Bedu, having lived with the Al Murra people of Saudi Arabia for many months back in 1983. A fluent Arabic-speaker, with a command of several Bedu dialects, Roger headed the group interested in the people of the Sands, their life and their future, representing the third component of the project – for there is a wind of change brought on by the increased oil-wealth the country now enjoys and this is affecting life in the Sands.

Dhiyab's reputation as the European Bedu had gone before him. It hadn't taken Said Jabber and others long before they had given Roger the nickname Dhiyab, meaning 'the wolf', a term of respect, although with Roger's lupine features, we all thought he bore a passing resemblance to one.

Being readily accepted into the Bedu community was important for the project if we were to understand at first hand the changes taking place. Dhiyab, who had been able to get a first-hand look at the Sands during the mapping phase, had chosen a team to help him focus on two main research topics: first to get some idea of what kind of people are living in and around the Sands and how they derive a livelihood; and secondly what changes might be taking place as a result of the increased wealth which is influencing the whole of the Badiyah area.

These aims neatly tied into the overall focus of the project and indeed would require the information about the plants and dune stability that would be provided by Paul Munton's and Andrew Warren's teams. Just as Andrew and Paul had different approaches and techniques, so did Roger. There are less rules in the social sciences, particularly if you are a Bedu, which Roger certainly is at heart. To grasp the whole picture requires discussion – a great deal of it – to get to grips with the complex social and traditional skills handed down by generations that influence the life of the Bedu today. Dhiyab was popular with the Bedu and he formed a close relationship with Said Jabber that will last a lifetime.

Dhiyab was the least outspoken of the three programme directors but none the less forceful in his views. Often he would carefully consider the discussions taking place at our planning meetings and after a number of puffs on his pipe would gently lay his opinions before us. These were always much respected and a natural balance to the campaigns of Andrew and the visions of Paul. Indeed, this was one of the strengths of a multi-disciplinary team.

Certainly his approach in the field was different. A notebook and a bag of dates were the tools of his trade. Discreet movements around the Sands and plenty of time to meet his fellow informants were to be the key to his success. This was a marked contrast to the high-tech approach of the earth scientists, but there was never a word of complaint. He and his team understood the different techniques and respected the patience and fortitude required to cope with technological breakdowns. On only one occasion did I have to give Roger my assurance that we thought as much of the human sciences team as the others for there were times when he thought they were being forgotten. How wrong he was. Remarkably, his team was able to establish a dialogue with a network of over 3,000 Bedu informants whose traditional knowledge was to enable Roger to get a detailed picture of the tribes in and around the Sands.

The United Nations Educational, Social and Cultural Organization (UNESCO) set up a Man and the Biosphere Programme to consider how man can better manage his environment to the benefit of himself and all the living things that share it with him. After ten years' work a major recommendation emerged. This was that there needed to be better communication and dialogue between three significant groups of people, forming a development triangle. The first group are those decision-makers and administrators who sit behind desks drawing up policies, recommending strategies and implementing expenditure of funds. The second group comprises those who are responsible for collecting geographical information, scientific or otherwise, about a region or ecosystem, and whose advice the first group should be incorporating into long-term plans. The third, and often neglected, group are those who live in the area under question and whose intimate knowledge is seldom tapped. These are the guardians and caretakers of the region and their co-operation and opinions should be part of the decision-making process, and thereby future development schemes, or they are doomed to fail from day one. Once there is a three-way dialogue which includes frank and honest comment about the priorities of each group, then it is generally agreed that there is a platform of information from which to build a system that can cope with the economic and social pressures of development.

Dhiyab and his team were continually reminding us that we must balance our findings with both the current needs of the people of the Sands and the knowledge they have accumulated over hundreds of years. His team, a co-ordinated and committed group of Arabists, wanted to get to the core of Bedu life and identify its structure and the vision of the participants. These masters of the Sands have decided to retain a Bedouin existence. It is their choice and one much respected by other tribes. This has a profound effect on the ecosystem. High on the agenda of Dhiyab's team was an understanding of the interrelationships of the three main economies – agriculture, pastoralism and fishing.

It is for this reason the notebook approach of this team complemented the technology of the earth scientists. When you have both satellites and Bedu on your side there is no limit to the data you might collect, so long as there is a form of communication between the two. The Crown Prince of Jordan endorsed this approach in a recent report on desertification by writing 'There is an urgent need to submit this precious knowledge and ensuing practices to modern scientific analysis and to organize a continuous dialogue between scientists, administrators, farmers and herdsmen.' This is the very stuff of the project. That dialogue was in the hands of Roger and his team, and it is a responsibility they took on board with incredible energy. I would not wish to go exploring without such a team ever again, for it is they who reached the spirit of the people and enabled us to hear the heartbeat of the Sands.

I remember Wilfred Thesiger referring to this powerful combination of innate and empirical knowledge found in tribal communities. Survival is not the sole aim of the Bedu; there is also a co-ordinated effort not to undercut the ecological basis of the desert upon which their future welfare depends. The Bedu have their own instinctive management plans and are careful not to overgraze areas or fell too many prosopis trees. This inbuilt control of resources was never more apparent than after rains when, in the first few weeks of rapid growth, seedlings needed all the help they could get. The Bedu would expend considerable energies on moving their herds to alternative pastures to allow the plants to survive. This was desert transhumance at its best and it is understood by even Bedu children. Such respect for the natural system has, in the case of the Sands, ensured that the full tolerance of the environmental network has yet to be tested; the Bedu have retained sufficient self-regulating systems to adopt, adapt and improve. However, objects of development such as the pick-up truck are now putting increased pressures on the environmentally balanced life of the Bedu.

Our brief was to try to identify man's effect on natural resources with the ultimate objective of conserving the best of the old traditions of resource use whilst recommending the most beneficial novel uses and developments. The project would then be in a position to share this information with those in Muscat – the 'decision-makers' – so making the third side of the development triangle. Roger, Paul and Andrew were the ideal people to draw that triangle.

While Roger didn't need expensive technological tools, he did require the skills of guides and interpreters. The six Omanis who consequently joined the team acted as go-betweens and introduced us to the main body of informants. Like Said Jabber, all became companions and added to the conviviality of the team. Whenever there were Omanis around, our people always seemed cheerful and in good spirits. It was always a pleasure to work with them and, unlike our own team, they never seemed to have low days. I had to bear in mind the advice offered by both the Wali and his deputy concerning who should and who shouldn't work for us. We were politely reminded we must have representatives of both the Al Wahibah and the Al Hagari so we chose one camp guard from each. Said Jabber was of course a pillar of strength; Dhiyab and he got along famously and it was through Said and his sister Latifa that Dhiyab was introduced to the majority of the pastoralists along the west side of the Sands.

During the early days of the project we had begun to meet a number of local farmers and traders who wished to offer us their services. I was woken early one morning, even before the call for morning tea had acted as a gong, by the sound of one of the guards shouting my name with some urgency. I bounded out of my room and out into the couryard, there to meet a man with one of the largest grins I had ever seen and an open, friendly face with extremely mischievous eyes. This was Khalifa Diwain Hilays Al Wahibi, agriculturalist, pastoralist, fisherman, trader, craftsman and comic. We shook hands; his was hard and work-roughened and I discreetly wiped off a resultant piece of fish slime from mine. He spoke good English and laughed at my attempt to greet him in Arabic.

'I am sorry I smell so bad,' he said. 'I have just come from the coast with my fish. Do you want some? Why are you not up yet – are you poorly?'

So began our relationship with Khalifa. His generosity never ceased and nor did the telltale smells of his activities. He was a fish trader and owned a pick-up truck, which was now full of his fresh and not so fresh catch. His relations fished on the coast and he had a house in Mintirib. We soon learnt that Khalifa in fact had relations everywhere and his range of contacts was impressive; he was certainly an excellent informant. He was keen to earn some extra cash, although I suspect he approached the project because he simply liked being with people and thought that we would be entertaining companionship for a while.

Khalifa had seen a great deal of life in changing circumstances. Married three times and divorced twice, he laughingly explained that it was because of 'too much yak yak yak'. He had spent his early years on Masirah Island, where he had become a fisherman and eventually the owner of a large boat. A shortage of crew members eventually forced him to move on in search of fresh opportunities. These he found, firstly with the British Royal Air Force based in Masirah Island and then with a wide variety of commercial companies which were mostly involved in road construction and transportation. Between jobs, Khalifa referred to his employment status as 'on-standby' and he was 'on-standby' when he joined up with us. He had just managed to keep the wolf from the door, providing for his young wife and daughter by his trade in fish. This involved travelling to his sister's house in Dhukhwayr, a small village on the coast, collecting ice on the way, and once there buying as much fish as possible. This he transported to market either to Mintirib, Ibra or even as far as the United Arab Emirates. His skill and knowledge as a fisherman were useful assets and occasionally he would go out on the boats as a crew member and receive a share of the catch in payment. His time with the RAF on Masirah Island had left him with some colourful language and a repertoire of risqué songs to complete his perfect command of the English language.

Nearly all members of the project had dealings with Khalifa at some time or other and the infectious laugh that was his hallmark will be fondly remembered. It was Angela Christie who got to know him best of all, often travelling in his fish-laden truck. Khalifa had a wicked sense of humour. On one occasion he was taking Angela and others interested in the fishing communities over to Masirah Island in a small boat. The sea was a little choppy and to divert attention from it Khalifa explained that it was traditional for those nearing Masirah for the first time to warn the spirits there by making braying noises much like a donkey might. Angela couldn't believe her ears but, not wanting to cause offence, began to make 'eeyor, eeyor' noises as they neared the shore. In fact, the whole boatload joined together into one almighty bray, with necks straining and faces turned up to the sky. Only when Angela caught sight of Khalifa at the back of the boat, holding his sides and with tears streaming down his face, did she realize that this form of arrival was not so traditional.

Soon after our introduction to Khalifa, the Deputy Na'ib Wali paid a courtesy visit to Taylorbase. We settled down to *qahwa* and dates in the *majlis* and discussed our progress. Yes, both Ali and Salim were being excellent guards and we had no worry about the safety of the camp. No, we had not had any further disturbances. The Deputy Wali's concern for our welfare was comforting. Disputes between members of a project and the local inhabitants are not uncommon and it was something for which I was keen to keep a look out. There were still rumblings over the use of Wahiba as a description of the Sands, although our new notice board referred to the area as the Eastern Region Sands. Should such a dispute arise, it was essential that the government representatives understood the project and knew its members. For instance, when one of our team accidentally ran over and killed a prize goat on M2, it was the intervention of the Na'ib Wali that prevented acrimony and a price of Omani Rials 70 (£160) was agreed as compensation. Everyone was happy except the goat.

The Deputy Wali was keen to help the project and asked why we didn't take on another English-speaking guide. When I enquired if he had anyone in mind he immediately suggested a young chap called Abdulhakim bin Amor bin Nasser Al Yahamadi from Shariq.

Abdulhakim arrived early the next morning with a snow-white *dish-dasha*, a waft of expensive perfume and a copy of the *Oman Observer*. His handshake was soft and gentle. Here was the town representative. I paused for two seconds while I considered the financial implications of yet more staff but, remembering the highest recommendation from the Na'ib Wali,

I welcomed Abdulhakim on to the project. Abdulhakim said he didn't want money but simply wanted to work with members of the project to improve his English. As he was reserved and slightly shy I was unsure what he could do for the team.

As Abdulhakim left to return to his home, he handed me the newspaper with an article about the project in it. This was the first copy of the daily paper that Abdulhakim brought each morning. His generosity matched Khalifa's and, in a similar fashion, he became attached to the project in very many ways. Abdulhakim, although a paid member of staff, was more a member of the team and joined in all our meetings and discussions accordingly. While he mainly helped Dhiyab's team he became a general adviser to all the members, who were keen to learn more of the Badiyah area. On several occasions he acted as a guide and interpreter for me, particularly when I was showing visitors around. I remember that on one happy day when Richard and Elizabeth Dalton from the British Embassy visited the Sands, it was Abdulhakim who made sure they were well looked after by local people.

It was during this time that we visited an extraordinary private library, quietly tucked away in one of the houses in the middle of Mintirib. It is one of the better-known houses in the town and a visitor simply has to ask for directions in the *suuq*. On entering, the heat and dust of Arabia give way to a modern, neatly organized library of books in Arabic that would be the envy of any university. It had been started by the grandfather of the current owner who kindly showed us around, having first offered us delicious dates. Mindful to clean our hands carefully, we were taken into the main room which contained thousands of books, references, maps and valuable copies of the Koran. One hand-written version was some 400 years old, but in pristine condition, carefully stored and catalogued in an air-conditioned room. No expense had been spared and there were wooden panels, discreet lighting and large tables at which to work. The owner was clearly proud of it and welcomed any scholars keen to undertake research.

Abdulhakim's generosity and openness meant we learnt a great deal about the Badiyah area. He treated us as his extended family and we were made welcome at his home at any time, day or night. His father was a prominent businessman and landowner on Shariq and of course Abdulhakim knew everyone. His close association with the team, while improving his English, led to considerable sadness when we departed at the end of the project.

Said, Khalifa and Abdulhakim were a formidable crew. Between them, they were able to introduce us to anyone of note as well as direct us to any corner of the Sands or any house on the margins. By luck rather than judgement we had a perfect team and they all contributed to the dialogue that we wanted. The project was to leave its mark on each – a responsibility I was worried about but could do nothing to change.

There is no doubt that development in Oman has been rapid. Never has so much been created in so short a space of time. Even during the brief time I had known Oman there had been radical alteration to the capital, best shown by the ever-changing pattern of buildings and roads, hotels, a mile of magnificent government ministries, a new international airport, five hospitals, two stadiums, the vast garrison headquarters and the new Sultan Qaboos University. This change has been a direct result of increased oil-wealth, but has only been made possible by the commitment to modernization on the part of the Sultan and his advisers. While the Sultan has been mindful of retaining 'the best of the past for the best of the future', the effects of such rapid change have been profound in both social and economic terms, creating a sharp contrast between the traditional ways of life that have remained unchanged for hundreds of years and the modern ones which began in earnest in 1972.

For the $1\frac{1}{2}$ million people living in Oman, this has meant coming to terms with many influences from the rest of the developing and developed world. This wind of change has passed through every village throughout the Sultanate and has been incorporated as a natural evolution of the country. While it is true that for some the wind is but a distant breeze, the visions and horizons for a new Oman age are nevertheless being discussed by youth and elders in all communities, in front of the television and around a Bedu fire. No one has had to focus more sharply on this than the Bedu of the Sands.

For Said Jabber, there was a continual dilemma as to whether he would bring his seven children up in the Sands and thereby retain the desert traditions and culture evolved over generations that are the backbone of his life, or whether he should live in Mintirib and send them all to the new schools in the Badiyah area. His compromise was to have a house in Mintirib and also to spend at least five months of the year with his sister Latifa, who moved her goats and camels around the Sands to where the vegetation was least parched. We could understand his predicament. His children would have to compete in the rapid development of the towns and villages at the north end of the Sands. The new Muscat–Sur road had ensured a permanent link with the capital on the other side of the Hajar mountains. The pace would not slacken and for a young boy, proud of his Bedu inheritance, there were many questions concerning his future role in the new Sultanate.

While changes in the villages were profound – many new buildings with air conditioning, mains electricity, televisions, telephones, supermarkets, new schools, new hospitals and so on – those for the Bedu were less marked. The bright red Toyota pick-up trucks so popular throughout the rural communities and the plastic thermos flask were two practical tools that epitomized the way in which the new was being integrated into desert life. The truck gave Said and his family much greater freedom in moving water, wood, food and goats throughout the Sands. The copper coffee pot, the traditional symbol of Arabian hospitality which was originated by the Bedu, has been replaced by the flask that only needs filling once or twice a day, thereby avoiding constant and expensive fires.

For Dhiyab and his team of Arabists there were two challenges. The first was to be accepted by the communities in and around the Sands and to be allowed to pry into their respective activities to see at first hand how the changes are influencing family life. The second was to be able to assess the longer-term changes by simply having seen the wide arena of activity through a relatively small window. The agriculturalists, pastoralists and fishermen of the area, all interlinked in the process of socio-economic change, had separate priorities and values. In accepting these challenges Dhiyab became responsible for promoting the future visions of each. For this he had brought together a number of specialists, all of whom could make a specific contribution to the monitoring of the changes which would be valuable to those responsible for the healthy long-term development of the Sands. In this process of consultation the Bedu guardians were important informants concerning indigenous industries including pastoralism, fishing, farming and craft work.

Khalifa's good friend Angela Christie from Durham University, a girl with sparkling blue eyes and striking fair hair, was the most popular with the Bedu. As with all the women members, she was given an 'honorary male' status when invited into the homes of those with whom she travelled. Khalifa nicknamed her Mr Angular and that was it. Mr Angular she became, although she was unable to hide her femininity even when laden with dust and grime from several weeks of travelling with the Bedu. Mr Angular was interested in how the commodities of the area, including fruit, vegetables, goats, camels, fish and locally made crafts such as camel saddles and blankets, could support the region's growing population. Studying the marketing strategies that ensure the producers can sell these at realistic prices to the consumers, either through the local markets at Mintirib and Sanaw or elsewhere in the Sultanate, was to be her primary task, particularly to see if there were any bottle-necks in the processes that could be identified and then ironed out. This she did by visiting these markets personally, talking to those who buy and sell produce and then meeting government representatives from the Ministry of Agriculture.

From Angela's point of view Khalifa was a valuable assistant; her work would have suffered without his help and his truck, in which they hurtled around the Sands. He knew the Sands intimately and was well-liked by those living there; he was recognized everywhere and constantly welcomed into homes where he would exchange news, adding vitality and loud laughter to any social gathering. With Khalifa, Angela was able to meet and discuss marketing and farming issues with all those they met. The trust that was established between the two enabled Angela to become 'one of the family'.

During her work Angela was able to attend a ceremony known as the Zar, held in order to heal someone possessed of an evil spirit. On this occasion the victim was a woman, her symptom being a severe headache. The Zar lasted several days during which there was singing, dancing and drumming, usually in the evenings. The Zar is considered to be a powerful and effective healing tool and the Zar leaders are treated with immense respect. During the Zar, friends and relatives of the sick person sit in a large circle, four or five men drumming, women sitting to the left of them and men to the right. The rhythm of the Zar often went on all night.

Around the Sands there are six main market towns, all of which are important outlets and well worth visiting on market day. The *suuq* is always easy to find, as it is a focal point of a town and is situated close to the mosque. Ibra, Sanaw, Mintirib, Kamil, Bilad Bani Bu Ali and Bilad Bani Bu Hasan were familiar to all of Dhiyab's team by the end of the project and all our members enjoyed a visit to the *suuq* when having a rest from the Sands.

Trading within the Sharqiya is becoming an increasingly important economic activity and the numbers involved in it are growing. Sanaw *suuq* is the largest and is considered the economic capital of the region, including Mudhaybi, Aflaj and the Barzaman area. It is a large, covered square surrounded by shops selling foodstuffs, clothing and jewellery, also built in a square to give the whole market a fort-like appearance. Open daily and especially busy on Thursdays, this market is a bustle of auctioneers selling goats, camels, alfalfa for fodder, fish, dates and farm produce. The auctioneer is always a showman who commands the attention of those buying and selling by sometimes bursting into song, often rushing around banging the soft ground with his camel stick and always maintaining a steady flow of banter. On one occasion I witnessed the goats being auctioned. A large circle of interested parties was formed, some squatting on their haunches, some standing. The owners then paraded their goats around in a great hubbub of activity and noise. Each goat is auctioned in turn, being carefully inspected by prospective buyers to the accompaniment of a great deal of friendly heckling. No more than 15 or so are sold in a day; prices vary enormously but a well-bred goat, in good condition with plenty of active service left, may fetch as much as £200, although that is still rare. Usually the auction is all over by 10.00 a.m. and the goats are then herded back on to the red pick-ups.

It is the camels who make most of the fuss. A camel sitting on its haunches in the back of a pick-up truck is the closest juxtaposition of the old and new that can be seen throughout Arabia and always induces a wry smile in new arrivals. At the *suuq* the laborious process of enticing a camel which has been trussed up to get off the truck backwards can take up to half an hour of pushing, cajoling, beating and pulling the beast, which makes an unholy din, bellowing from deep inside and spitting a yellow phlegm on all those who get too close. The smell is memorable.

The other markets are smaller but still a focus for trade, although they are likely to open on just one day. Wednesday was market day in Mintirib, Thursday in Kamil, while Friday was busiest in the Bilads. The local farmer will sell direct to the covered *suuqs*, but is now beginning to compete with imported foodstuffs from the United Arab Emirates; the fisherman will deal through a trader such as Khalifa. As in markets worldwide, the social aspects of meeting and sharing information are fundamental. Good old-fashioned gossip is important and many Bedu will visit on market day simply to catch up with news of friends and family or of where the best grazing may be.

The three communities in the Sands occupy different niches. The pastoralists live within the Sands, predominantly in the western and eastern margins of the prosopis and on the northern high dunes, the farmers rely on the *falaj* irrigations of the oases of the north and the fishermen have established themselves on the coast although they migrate northwards to the towns in summer.

Predictably the sale of goats or camels is the least organized and established trade. The 3,000 or so Bedu have their roots in subsistence farming, only looking after their immediate needs. Selling to the markets contributes to their daily living requirements, but Angela found that nearly all Bedu have other significant sources of income such as a date garden in one of the villages or a family member in the services either in Muscat or even in the United Arab Emirates. They will travel regularly to the market, but there is still a degree of informality. A herder will generally not know how many animals he owns, how many he has sold or intends to or whether he will be able to afford the costs of looking after the herd during the severe drought periods, when the price of animal food becomes expensive. The traditional patterns of keeping goats and camels are retained within the community for social reasons rather than for pure economic incentives. The important issue for Dhiyab and his team was to determine whether pastoralism was on the decline or if there were signs it was evolving to become better integrated into the lives of those who only partially lived in the Sands.

Life for the pastoralist Bedu is still dictated by narrow margins of survival. When rains don't fall for three to four years the Sands become sparse of vegetation and life can become harsh. However, the Bedu live there by choice rather than because they have to and it is the nomadic existence that enables them to sustain a partnership with the slim but sustainable resources, even through periods of drought. Nevertheless, it is this nomadic lifestyle that presents the greatest dilemma to those parents concerned that their children should go to school. It was Dhiyab who gleaned that the Bedu wish to share in the process of modernization, but believe that haphazard, disconnected development will cause more problems in the short term than it solves in the long.

The pastoral population of the Sands is estimated at 3,000 persons or some 500 households, with the most densely populated areas being the prosopis woodlands bordering the Sands on the east and west. The Bedu of the Sands, who belong to several different tribes, are nomadic herders of goats and camels, with other livestock playing a minor role. They make use of the different zones and resources of the Sands at different seasons. Rainfall stimulates the Bedu to move from one area to another if it is sufficient to make new growth of ephemeral vegetation appear, or enough to make green the huge areas of *Panicum turgidum* grasslands in the west of the Sands. As it may rain only every three or four years such movements are infrequent, but when they occur they are dramatic with large herds being driven up to 90 kilometres. A long way on a camel, as I was to learn; not so in a pick-up.

The prosopis trees are important to the Bedu as a source of shade, browse, fuel and building wood. Further into the Sands there are no trees to offer shelter. However, the Sands do provide excellent grazing in dry periods and the perennial plants, known locally as *rimram* (*Heliotropium kotschyi*), *thidda* (*Cyperus*) and *'arta* (*Calligonum comosum*), are important. After the rains fell in February a number of annual plants immediately grew up and this grazing, although shortlived, is the best of all.

Dhiyab travelled extensively throughout the Sands on his own or with Said Jabber and I had the privilege of travelling with them towards the end of the project. Dhiyab's reputation as the European who could speak fluent Bedu Arabic went before him, which helped in his introduction to the many families living in the Sands. Consequently Dhiyab met more of the Sands people than the rest of us and through discussions late into the night he pieced together an overall picture of the landlords of the Sands.

There is a number of tribes living in the area that we call the Wahiba Sands. In each district or village one tribe is usually predominant but although tribes are associated with a particular area there are no precise or universally accepted boundaries and the tribes may mix together. However, over 95 per cent of those who live in the Sands belong to the

Al Wahibah tribe, which is further divided into subsections. A tribe comprises direct descendants of some remote ancestor whose name is often that of the tribe and, in this case, the tribe originated many generations ago from a woman named Wahibah. Said Jabber explained to me that the boundary of the Al Wahibah was that clear line drawn where Wadi Batha and the Sands meet. In fact, when we drove south one day he stopped and drew a line on the precise edge of the wadi and pointed with his camel stick. To the north are the Hajriyin, the mountain people, to the south the Al Wahibah – the Sands people. There is no law or prohibition, just a matter of tradition and preference.

Most of the Al Wahibah tribe will use Sanaw and the Badiyah towns for trade or even buy a house or date garden. The sub-families were divided into localities thus:

AL WAHIBAH	Khawasiyat – Wadi Murayr
– Bin Hayyah	North and central Sands
– Hal Anfarri	Southern Sands
– Hikman	South around Hajj, Jawbah
– Al'Amr	Wadi Sayl and along coast
– Yahahif	Wadis Halfayn and Andam

The Al Bu 'Isa is one of the smallest Bedu tribes and has close connections with the Al Wahibah. Traditionally found in the southern portion of the eastern prosopis woodlands southwards to the coast, incorporating some of the fishing communities, this tribe tends to use the town of Bilad Bani Bu Hasan.

The Janabah tribe was once considered the most powerful Bedu tribe in Oman and was traditionally at war with the Al Wahibah, although there is no evidence of this today. While the Janabah tend to base themselves in Bilad Bani Bu Ali, they are widespread in separate enclaves throughout central and eastern Oman. Few live in the Sands proper but are found in the eastern prosopis woodlands and along the coast north of Khuwaymah.

The Hajriyin or mountain tribe are a large group now found to the north of the Sands within the Badiyah area. They do not enter the Sands at all, while other tribes found in the north who occasionally do are the Mawalik tribe, based in the north west near Qabil, and the Hishm, from Kamil in the north east.

The Hajriyin or mountain tribe are a large group now found to the north of the Sands within the Badiyah area. They do not enter the Sands at all, while other tribes found in the north who occasionally do are the Mawalik tribe, based in the north west near Qabil, and the Hishm, from Kamil in the north east.

Tribalism is an important feature of day-to-day life. Members of the same tribe tend to live together and to co-operate economically. Among the nomadic groups, members of closely related segments often move together and many share some of the tasks of herding animals and supplying the camp, which is sometimes based on a core of brothers with their wives, children, cousins and grandparents. These extended families will operate independently and it is only at a time of crisis that the whole tribe will be called together to make joint decisions. Tribalism retains an important role in local government; those who want to influence the government do so with the support of their tribe and through their tribal leaders the Sheikhs. Today, most tribal leaders are making efforts to ensure that the younger members of their families receive an education, which many older Sheikhs lack, so that they will be able to fulfil their role of intermediary between tribesmen and government more effectively in the future.

The Sands has a long straight coastline where the blue Arabian Sea breaks on pale shell sands or low cliffs of pale rock, remains of the long immobilized dunes of past deserts. The seas bordering many deserts of the world are rich in life and this one is no exception: when flying over it we saw many turtles, rays, sharks, cetaceans and many schools of fish. William Lancaster lived on this coast with the Bedu fishermen who exploited this natural wealth. He was no stranger to living with Bedu, for William and his family had been accorded the rare privilege of being allowed to live for several years with the powerful and dynamic Ruwallah tribe in Saudi Arabia. He soon settled down in the house of an elderly man in one of the coastal villages.

The village was situated beneath imposing dunes which rose to about 100 metres and on hot days were only just passable by Land Rover from inland. The villages comprise a number of homes typical of those used by nomadic Bedu. They are movable structures consisting of an oblong wooden (or sometimes metal) frame, rounded above, over which thick woven goat hair would once have been placed; now canvas or tenting material is used. There is an added fence to keep out the very few livestock that live in the village and this may form an enclosure several times the size of the house. The lack of any vegetation whatsoever for several square kilometres around these coastal villages is striking. This is probably because large movements of sand by the south-west monsoon wind in summer snuff out any plants struggling to live there.

Each village has about fifteen houses apparently placed at random for there is no sense of a road or streets: the sand continually fills in tyre tracks and there are no rocks or vegetation to force trucks or pedestrians to take any particular route so no permanent tracks are created. There are also a few shelters from the sun consisting of a flat roof of dried date palm leaves lying on a frame supported by four simple wooden posts. In this shade a mat is placed, upon which people of different households or visitors sit talking and drinking coffee; when the people leave, taking the mat with them, the shade is quickly taken over by the goats. These and the houses are the only man-made structures besides the wells, each with a low round, slightly convex wall, narrower at the top than the base, made of grey cement and with two pieces of wood meeting to form an apex over the well mouth; from this is suspended a pulley to take the water bucket's rope. The bucket is made of a piece of tyre inner tube sown round with thick string which secures it to the rope. The wells are surprisingly shallow, good fresh water lying just below the surface in spite of the proximity of the sea. Without this water there would never have been any settlements on this harsh seaboard. The only other things in the village is a few empty 45-gallon petrol drums scattered at odd angles, and the inevitable pile of tin cans alternately buried and exposed by the shifting sand.

William's village was situated about a kilometre from the sea where the boats were to be found drawn up on the beach alongside damaged nets. Nearby shark and other fish dried in the sun, protected from seagulls and Ruppell's sand fox by old netting. The foxes are very common in this area, probably because the sea yields a rich flotsam from the marine life off the coast and jetsam from the fishing boats. Just above the seashore ran an ungraded track about half a kilometre wide in places where vehicles had used all the ground available to them to find the best surface and avoid patches where other vehicles had bogged in the soft sand.

William's main problem was boredom, because, in contrast to the dynamic politics and trade of the Ruwallah, almost nothing happened on this Omani sea shore apart from the milking of a goat in the early morning and, almost every other day, the departure at 4 o'clock in the morning of fishing boats and their return later in the day. The old man had very little to say and contacts with outsiders were few. This contact seemed to have been actively discouraged by the Bedu, withdrawing their village inland from the coastal track. In times of raiding and feuding the village was on the coast because of the necessity of guarding the boats or being able to flee out to sea if attack came from inland. The boredom was added to by featureless rolling dunes and a persistent light sea breeze which together amounted to a type of perceptual deprivation that has a debilitating effect on the minds of those exposed to it for long periods: it seemed to have resulted in a very special sort of lethargy in these coastal people not found elsewhere in the Sands.

The monotony of life was broken one day by a wedding. Then tens of Bedu trucks came struggling up from the inland side of the dunes to roar down the seaward side and park in haphazard mimicry of the disposition of the Bedu houses. Out stepped the men in cleanest white *dishdashas*, their figures emphasized by the belts decorated with silver wire clinging to their waists in which was thrust the curved Omani dagger in a silver sheath. On their heads were cloths in a multitude of colours each wound by its wearer in a subtly different way to display the cloth or accentuate the bearer's features. They strode over to where the men were gathered to make the long Bedu greetings that convey, by the order and type of greetings, both social standing and the minutiae of intra-tribal association. Young men who could not afford a dagger and belt nevertheless affected to swagger with a short hooked cane, the camel stick. The women emerged more slowly from the trucks, adjusting the dark overgarment to cover brighter cloth beneath, rewinding garments in the shelter of the trucks and checking their headdresses before they too walked slowly and dignified as the men, but to the place where the women all sat, masked and in a row, to the stranger each indistinguishable from the next. Food, young goat, rice and nuts were being prepared behind a canvas screen erected on poles to exclude the sand so easily blown across the round shoulders of the dunes. Drumming started and the men ranged up in white rows to dance, watched by black rows of seated women. Nearby in the cool shade of one of the houses, the bride of about 14 years sat, out of sight of men, waiting to be taken to the bridegroom's house.

Fishing is profitable and the people could have made a lot of money from the fish they caught but seemed uninterested in the prospect of expanding their fishing activity; they apparently have sufficient for their needs, although no household had an electrical generator so there were no televisions or other electrical goods and few radios. They were, however, interested in the idea of a proper road and especially keen for a health clinic to replace the infrequent visits of a medical official. They were ambiguous about formal education, fearing loss of basic fishing skills. Unusually for Oman the young men had stayed in the village either to fish or to earn money as crew on larger fishing boats. They merely took their catches into Nuqdah or other larger villages with better roads close by where it was bought by merchants who carried it with ice from the local ice factory to be sold in the capital area or further north as far as Saudi Arabia or, some said, Jordan.

The life of these fishermen challenged the project members' views of the nature of Bedu. Although they only had enough goats to eat household scraps, and their winter income was entirely from fish, they regarded themselves as Bedu and were clearly regarded as such by other Bedu. In the summer they move to towns in the north of the Sands, apparently because the south-west monsoon makes fishing in small boats impossible and because the wind drives the sand, making life difficult. In towns they contract themselves out as labour to bring in the date harvest.

The link of the fishing Bedu to the towns 100 kilometres or more to the north is one example of the complex interrelations between pastoralists, fishermen and the townspeople (who are traditionally gardeners or agriculturalists). These divisions were also reflected in the structure of the project because when in the summer the fishing Bedu leave the area studied by William Lancaster, they move into the area where Tom Gabriel was studying the gardens to the north of the Sands.

After hours walking across soft-yielding pale sands where the light is so bright that it is hardly possible to open your eyes, and the sweat runs down the forehead and back, it is a very special sort of bliss to enter a palm garden. Here the coolness envelops one, the eyes open to see the palms rising upwards like pillars, each supporting an umbrella of dark leaves yielding a dappled shade, the ear is lulled by the sound of rushing water, and there is green grass to rest upon. Such contrast is the basis of the Koran's concept of Paradise, literally translated as the Garden. For those that fear God there are two gardens where they will lie on couches of thick green brocade, within their reach will hang the fruits of trees, the pomegranate and the palm, and gushing fountains and rolling streams flow there. There the believers will live with bashful virgins, untouched by man or *jinn*, fair as hidden pearls.

But the reality of the earthly paradise is that the management of these gardens is a highly political and social matter. This is because these gardens are watered by a system called the *falaj*. This is a system of water channels, some subterranean and others open conduits, which bring water to the gardens, often from springs many miles away. The slopes are very gently graded so that water runs the maximum possible distance for the slope available: there are clever syphon mechanisms which allow the water to negotiate wadis, passing down the sides, under the wadi bed and up the other side to resume its course. This protects the water channel from destruction during periods of wadi flood. These systems are technical triumphs of water engineering and their repair is undertaken by members of a local tribe whose personal gardens are full of miniature syphons and other feats of engineering.

Unfortunately the *falaj* is relatively costly to maintain and necessarily involves many people in paying for and organizing its maintenance and use. A gardener who is the final user of the water has to negotiate with many people and is dependent upon the consent of all other users. As a result the *falaj* gardener tends to be less flexible than the man with his own well from which he pumps water. Consequently Tom found that the reaction of *falaj* food producers to changes in markets and in the need for different sorts and quantities of produce is slower. The disadvantage of the well-based system is that there is no control over use of limited water in the watertable, and such uncontrolled exploitation of a limited resource has led in other parts of Arabia, including Oman, to depletion and salination of the water supply with devastating results for gardens and the gardeners. Most *falaj* systems, on the other hand, do not take more than the spring gives: this may decline after three years with substantial rain.

This was the situation that Tom Gabriel studied and he found a number of changes were in progress. There was considerable loss of young men to the capital area and to the UAE so that agricultural labour was being lost and young people were not learning agricultural skills. On the other hand a change from a purely subsistence agricultural orientation was taking place. There was a particular demand for fodder such as alfalfa and the local people had responded to this but there was no parallel grass production. On the other hand there was a need for more advice on growing better quality produce for human consumption so that it could compete with that from elsewhere in Oman and the United Arab Emirates and be exported from the area and fetch a better price.

The changes in the economy around the Sands have been symbolized by the materialization of the entrepreneur, a process speeded by government encouragement of enterprise. They symbolize the change from the traditional subsistence culture, involving perhaps only the export of dates, to a complex trading system. In her work on this group of people, Corien Hoek had speeded from one entrepreneur to another in the villages around the Sands and slowly began to realize the important part that this group played in the introduction of new technology and ideas. The role of a seller of food mixers, washing machines or hi-fi equipment is not only selling things; he is also educating the members of the local population in their existence, their relevance and how they work. The shops on the tarmac road run by immigrant Indian labour are futuristic exhibitions. Corien found that most entrepreneurs, who could be either men or women, ran a number of different concerns, the diversity giving stability to their enterprise. Bank loans have been of major importance in getting business moving. Building, including the making of cement bricks and the growing of alfalfa, are two favoured spheres of business responding to new needs.

Corien forsees the importance of such entrepreneurs growing and suggested that this would be strengthened by broader application of grants and loans and education of people in an awareness of business opportunities. Even so the new entrepreneurial spirit has its limitations for a trader, who will give cheap prices to his relatives or even give away goods free, so fulfilling traditional obligations to look after the welfare of his kin. This has a special twist where a trader has sold the right to run his business to an Asian expatriate, usually a Pakistani or Bangladeshi, for the payment of a regular fee. The expatriate has to bear the loss of the owner's relatives coming in and helping themselves!

In contrast to the study of the entrepreneur, Gigi Crocker and Charlotte

Heath toured the Sands talking to women and men about the traditional weaving crafts which use goat hair, sheep wool and palm leaves. They studied the techniques of producing a variety of things where they could find exponents of the skill. They visited many remote Bedu homes in the Sands and dark shops in obscure streets in nearby villages and towns where exponents of ancient crafts still worked and sold to a small discerning Omani clientele. They were able to get information about diverse articles: large rugs called *abya* constructed of a twined weft; the *durri*, a large basket with lid made of coiled strips of woven palm fronds and covered in leather which is used by the women for storage of their personal possessions; the man's embroidered hat, the *khumar*; camel trappings including the *mahowie*, a woven cloth which is attached to the saddle and surrounds the camel's hump; girth straps and saddlebags. An important traditional product is the strips of goat hair that were used to make walls on the frame of Bedu houses but which is rarely seen nowadays. There are some surprising indigenous local garments, the most striking of which is the sand sock, a thick dark sock made by looping, which provides the Bedu with protection against hot sand in the day and scorpions at night. These and the camel trappings which adorn the many racing camels of the area are still being produced, but other garments and mats are no longer being made because it is cheaper to buy mass-produced plastic mats and Indian cotton and ready made decorations for women's dresses than it is to make them. People still have a feeling for fine locally produced work so Gigi and Charlotte found that there was a market for what was made. Woven articles are still made and sold for high status activities like camel racing where various parts of the camel's harness is beautifully crafted. The very best weaving is reserved for the immediate family.

These two project members provided a rare and important insight into the demanding life of women in the Sands. Women are expected to herd goats or work in the gardens, carry water, cook and prepare food as well as produce and look after children. They may also weave and make clothes, carpets and camel dressings for the family. The lack of female doctors and teachers, in spite of the government's positive view towards both women's education and the broadening of their expectations, is indicative of the limited resources available to them in the Sands and the great difficulties of lightening the exhausting load they are expected to bear.

In the final days of the project the whole team took up the invitation to a feast at Aflaj, the home of Sheikh Mohammed bin Hamad bin Daghmal Al Wahibi, the head of the Al Wahibah tribe. Sheikh Mohammed was an elderly, stoutly built Bedu whose hospitality was legendary. This was a memorable occasion, for which five or six goats had been killed. A large number of Bedu had gathered, all in their finest *dishdashas* and sporting *khanjars* and varying colours of *masarra* (headscarves). Beside the open, acacia-strewn wadi, the Sheikh's house, thick-walled with a green roof, was built with a courtyard all the way round. Said and Dhiyab kept close to interpret and to help me with the protocol of having lunch with the paramount chief of the Sands.

We were all ushered into his *majlis,* leaving our shoes in a large pile outside. As we entered we were introduced to a number of the elders and there was much shaking of hands before we were all in and standing in a circle around the carpeted room, which had no furniture. There were a number of familiar faces that we had come to know over the past four months. We all waited for Sheikh Mohammed to settle down on the carpet, sitting down ourselves cross-legged, with one foot tucked under. Most of the team had become quite supple by now and could sit Bedu-style for long periods, but others found this position difficult to hold for long and had to keep shuffling to get comfortable.

Sheikh Mohammed and I exchanged further greetings. He enquired how the project had gone and I was pleased to explain how much we had seen and learnt, largely through the help and co-operation of the many Bedu who had become known to the project – and I made a particular point of thanking him for the special help that Said Jabber had given us. I handed the Sheikh an aerial photograph of his house that we had taken on one of the SAF reconnaissance flights and he recalled the plane flying overhead. The question of the project's notice board came up and we laughed nervously, concerned we had made a further faux pas, but no – he was simply interested in who else might be coming to study the Sands. Links with the university would obviously involve those who could continue to act as guides and field assistants. Sheikh Mohammed confirmed that his people had enjoyed the Royal Geographical Society's visit and offered continued help to those who would return in the future.

While we discussed the future, a feast the like of which we had not seen during our time in Oman was laid before us – dates, fruit and coffee followed by a huge meal of some eight large trays each a metre across filled with goat and rice. The goats had been cooking for some 24 hours in the traditional way and they were tender and delicious. Together with our Bedu companions, we shared the best the land had to offer, gorging ourselves on handfuls of rice and meat which we neatly rolled into little balls which popped not so neatly into our mouths. For a brief period, hosts and guests were lost together in the traditional and time-honoured sharing of food.

Later, standing outside at the end of the afternoon, we bid farewell to our companions. As we did so, we realized that, while the Al Wahibah are the guardians of the Sands, we had the responsibility of being their ambassadors.

CHAPTER SEVEN

The Wahiba Bear

Ali bin Abdullah bin Ha'atrush Al Wahibi gently scraped away the surface of a patch of sand near the camp. It was damp and formed a layer some 7.5 centimetres deep while the dry sand below slipped through his fingers. The sun had yet to rise and the unfamiliar morning freshness was supported by a cacophony of insect noise. 'Alhamdulilah', muttered Ali as he raised his arms outwards. This was evidence of heavy rain in the night, the first significant rainfall for many years. Some say three, some say more.

To date this was the most significant event of the project. Measured at 'half a hand deep', it is unlikely to induce an immediate response from the vegetation but Ali, who has a daily view over the northern end of the Sands from his position as camp guard and nightwatchman, thought there would be more to follow 'Inshalah'. During February there were to be a number of large thunderstorms in the Sultanate, many of which could be heard from Taylorbase, grumbling in the distance with sharp flashes of lightning every now and then. We soon realized that the wadis would be full with water flowing from the mountains. To see a full wadi is both frightening and impressive.

The Wadi Batha had risen dramatically when we went to inspect it by the light of our Land Rover headlights; we saw a sinister swirling mass of very dirty, sediment-laden water rushing past the north face of the Sands on its way to the coast. Even for an experienced geomorphologist like Ewan Anderson this was a memorable event. The pattern of short, unconnected lengths of wadi, depicted on the map of the area beyond the eastern edge of the Sand Sea, was immediately clarified. Wadi Batha burst its banks and part of its flow was deflected southwards, evidence that this was once its original course. The flow was supplemented by large inflows from the gravel terrace, and for a day Wadi Batha flowed in two directions at the same time. It coursed southwards along the edge of the Wahiba in broad anatomizing channels for 45 kilometres before arriving at a low sand ridge, just short of the final *sebkha* and its outlet to the ocean. There, its level rose, but it had insufficient strength to force a way through. Twenty-four hours later, the entire southern course was virtually dry. It is a rare experience to see the birth and death of a river, all in one day.

Judith Maizels and Chris McBean working at Wadi Andam near the pink cliffs of Barzaman saw dark thunder clouds and pitched their tent. News from Taylorbase confirmed that storms were on their way, but it was not until 2.00 a.m. that they heard a strange rushing sound and went out to investigate, only to be confronted by a wall of water passing just 50 metres away from their tent. Beck and Mike Holman, driving back from Field Base earlier in the day, had met a tidal wave coming down the track towards them and were soon engulfed by a torrent of water. As the water rapidly swept across the surrounding landscape following the path of least resistance, they tried to head for home. Soon only patches of high ground with bushes and a few scattered Bedu dwellings were visible above the water. It was difficult to tell the depth of the muddy water and they often had to get out of the vehicle to check their route. On driving through Bani Bu Ali, Mike found the main street a river with cars abandoned on all sides. Desperate to get back to Taylorbase to be on hand to help others who might be stuck, he bravely took his vehicle through the waters, often lapping over the bonnet.

The rain gave us the lucky opportunity to witness the process of keeping the Sands at bay, as the sand which had strayed into the path of the wadi was now borne seawards. The waters subsided as fast as they came, leaving a layer of mud and isolated pools. The boundary of the sand was now so sharp that your foot could be placed half on sand and half on gravel. Significantly the sands absorb all the water that falls; and the sands act like a sponge, holding the water. The top surface dries and forms a layer which retains the water, and this increased moisture gives the sands a firmness.

These substantial February rains fell all over the Sands but predominantly in the northern areas, and for the remaining two months of the project we were to see how the Bedu moved their herds away from key grazing areas to allow the grass to grow before they brought them back in. Below the prosopis trees at Field Base grew a carpet of grass 6–7 centimetres long, and the trees themselves, partly because the dust had been washed off and partly because they had a burst of new life, turned from a dull grey to a dark green. The rains brought many advantages to the team, not least the opportunity to see the changes in biological activity.

In the eastern part of the Sands, between the track we called the M3 and the Wadi Batha, the regular formation of swales and crests give way to a great jumble of sand dunes, small mountains of sand 500–600 metres high. It has been described as the Wahiba's own 'Empty Quarter'. Tracks were difficult to follow and the route-marking team had not had time to go through this area at the start of the project. Quite rightly, we had been apprehensive about operating freely in such areas; the idea of regularly seeking overturned Land Rovers amidst this soft sand was a nightmare I wished to avoid. On the main operations map at Taylorbase this area was marked 'Out of Bounds' in thick blue ink, and was a constant source of fascination to all who visited us.

During our four months in the field, I didn't think we were going to have time or resources to travel through the no-go area. It remained a low-priority task for the project until the young earth scientists, Robert Allison and Chris McBean, needed to work there collecting samples and ground truth. Both Bob and Chris, quite rightly, felt that their sampling of the Sands would not be complete until they had run a transect from north to south. They put this idea to Kriss. Kriss is only happy when he has a challenge on his hands and he set-to planning the logistics of the journey as soon as it had been mooted. The recent rains would help enormously. The ground would be very much firmer and would make it much easier to ride over the softer sand areas.

Kriss knew the area by reputation, although he had never traversed it from the Wadi Batha in the north right down to the coast in the South near Khumwaymah. He had seen enormous mounds of sand from the air and it looked pretty impassable country, despite never being more than 40 kilometres wide. He had been close to some of these mounds during a west–east crossing of the Sands that he had made with Tim Cheeter from the Desert Regiment. They did the crossing in the height of the summer when it was so hot that skin from the palms of their hands got stuck on the steering wheel when they forgot to wear gloves.

Of course the Bedu knew the area as the western part offered better grazing. If there was an area in the Sands still to be explored, this was it. It would be a good opportunity to visit an area where 'only Bedu had been before' as Chris and Bob would say. They clocked up an enormous mileage around the Sands in their quest to sample all its different locations. Working as a team, these two were happiest in the field though this was the first time Bob had left the UK. Their work took them to every corner of the Wahiba Sands as they had to take 400 sand samples, using Fablon to collect the top layer, and special readings to measure how bright the sands were. Chris had designed a special frame to help him take light reflectance readings of the Sands. These readings were to be used to calibrate or 'ground-truth' the information being collected by the satellites overhead.

These images of the world taken continually by orbiting satellites, when presented on paper, represent a series of coloured rectangles, which when put together form an overall picture of an area, similar to an artist's impression of the image seen by the compound eye of insects. In the image of the area around Taylorbase, the camp itself appears as an orange rectangle. The area covered by such a rectangle is known as a pixel and is an area of about 30 metres square. Chris's task was to take light reflectance readings within specific pixels to calibrate the overall satellite image of the Sands. The information collected by Bob concerning the mineral content, shape and size of the sand complemented Chris's data.

At each location Chris made a detailed description of the colour of the sand, its texture and vegetation cover. Then he would set out his odd looking aerial, which was in fact a radiometer on an aluminium pole, which measured the amount of light being reflected off the Sands. Readings are taken each through different filters – green, red and infrared – and the resultant figure is known as a spectral signature. Similar types of terrain should give similar spectral signatures, so in theory it is possible to correlate the spectral signatures on the satellite image with the information collected on the ground and extrapolate these conclusions to provide information on large areas. Such techniques are particularly useful for producing vegetation and water resource maps. The prosopis woodlands and wadis could be clearly seen on the images of the Sands.

Through the American space agency NASA, Chris had ordered a special image of the Sands to be taken during the project by a satellite known as Landsat TM. One single 'scene' taken by this satellite would cover the whole of the Sands. It comprises many 'spectral signatures' which, when compiled, form a picture similar to a photograph. Different colours represent different geographical features and give an overall indication of the amount of vegetation cover and soil moisture. But 'sands of a thousand colours' can play havoc with such readings. To cope with such diversity and incorporate Bob's information about the Sands, it was useful to get to as many locations as possible. The journey through the no-go area would add considerable data to their figures. I put my name down as a driver.

Kriss organized a convoy of two open-topped desert vehicles and early one morning we entered the Sands at a point off the Wadi Batha some 30 kilometres east of Taylorbase. Behind, in a wide arc, lay the rugged Hajar mountains in colours of mauve and pastel blue. To the south the sand disappeared into a shimmering horizon. This journey lasted five days and was not without its adventures. It taught me a great deal about the commitment of field scientists who go to extremes to collect their data.

Kriss led the convoy. His experience enabled us to pick a route through the dunes. At times we had to drive up very steep sides and then coast down the slipfaces. Being careful not to get stuck in hollows or drive diagonally across any steep hill, we attacked the crests with vengeance and then in the swales meandered through the *Calligonum* bushes.

Heading due south using the sun and a small Silva compass, we maintained a steady bearing despite having to circle round some of the larger dunes. A shadow and a small mark on the dashboard helped Kriss keep a direct bearing on the sun. He stopped often to reconnoitre the next section and check through his binoculars. In the distance some 50 kilometres away, small mountains of sand could be seen. They looked menacing and very pink.

I followed Kriss with Chris and Bob riding shotgun; with Kriss were Hamid, his radio operator, and William Lancaster. It is a marvellous way to travel. The rushing wind cools the face and masks the true heat of the sun. An all-round view gives the Sands an impressive grandeur as the mountains in the north become less and less distinct. Chris and Bob wore their *masarras*. I wore a white topee-like hat with a solar-powered fan on the front to cool the face which had been given to me by an eccentric American friend in London. Although ashamed to be seen wearing it with Bedu around, it worked well, the little fan burring fast inches from my nose.

We stopped after 10 kilometres for Bob and Chris to take their samples. Out came the contraption and Fablon. Bob had to take a sample of both the crest and swale and he would rush off perpendicular to the vehicle, bag in hand. Kriss used the time to check his position, listen in on the radio net to see if he was needed and drink tea from his thermos flask. I would help Chris by holding his radiometer and pulling a bit of string to expose the sensor four times. These stops were a welcome break. Kriss would proudly mark his map and say we were 'here' and place a large blue cross in felt pen to confirm it.

We usually pushed on after half an hour or so. The distant dunes always looked a long way away and as the sun got higher in the sky, they began to lose their pinkiness and looked almost white, like a large ice-cream. Although the going was still relatively easy, we had to be careful: the main swales are not quite north/south at this part of the Sands, and so to cross from one mega-dune to another there were many smaller seif dunes to be crossed, many of which had some spectacular sand crests, beautiful and pristine yet a potential hazard to amateur sand drivers.

There is a special technique to getting over the steeper dunes that I copied from Kriss. Usually it was possible to take a run at the slope and be going as fast as possible, changing down the gears fast as one went up the slope. Usually there was some sort of ledge or flatter area just before the crest and it was possible to turn on to that, drive along for a short while and peer over the edge, before making a sharp turn and driving the vehicle directly at the crest. It was important to hit the top of the crest with sufficient speed to cut through the top edge of sand, which may be three to four feet high. If not, the vehicle would become well and truly wedged. But to hit the crest too fast may mean launching the vehicle into the air and increasing the chances of rolling the vehicle on the other side.

We had some close shaves. Kriss, unusually, got stuck on a particularly sharp ridge. All four wheels were spinning and off the ground while the vehicle was left see-sawing on the thin band of soft white sand. On one side was a narrow valley and a large clump of bushes. On the other was a very steep slope that was covered in clumps of vegetation, all the way down to the bottom of the main swale below. It must have been a drop of at least 50 metres. We couldn't use the tow rope, nor could we push him with our bumper. We dug the middle ridge of sand from under the vehicle, used the balloon jack to get the back of the vehicle up in the air and scooped quantities of sand back into the holes where the wheels had been. We dug under the front tyres until the whole vehicle began to tilt down the slope and then pushed like fury while Kriss revved and revved and like a jack-in-the-box leapt over the side and down the slipface.

Determined not to have a repeat performance, Bob and Chris egged me on. With a mighty roar we hit the crest, but it was too hard. With sand spraying everywhere the nose of the vehicle went on rising and leapt high into the air. Our cries of excitement began to turn to fear as we saw what a huge drop it was the other side and that we were out of control. I frantically pressed the brake pedal, but with all four wheels in the air naturally to no effect. We must have landed a good 9 metres down the slipface and bounced. A box of army rations in the back bounced even harder and flew up over our heads and down the hill in front of us. Kriss, William and Hamid were rocking with laughter. This was good training for the big dunes to come. We stopped for a brew even though it wasn't a 10 kilometre mark.

At each stop, it was noticeable how much wildlife there was. The influence of the February rains was very much in evidence. Most of the plants were in flower and telltale tracks in the sands hinted at a large population of nocturnal animals. In one swale we noticed a mass of caterpillars on the move, chomping at anything in their way. As we looked closer and followed the tracks of one or two of these brightly coloured creatures, Hamid pointed out grasshoppers, locusts, ichneumon flies, cicadas and the odd small lizard. We persuaded Hamid to give up his sweet tin so we could take some specimens back to Willie and Sonya. The predominant vegetation was varying sizes of *Calligonum* bushes in full flower, a blaze of bright red. A thin layer of green covered the whole of the valley and we concluded that there must have been even more rain over the past two or three weeks.

It was at this point that Kriss heard on the radio that Judith Maizels' vehicle would not start and that she was stuck on her own south of Barzaman. He radioed the army mechanics with their pick-up vehicle at Mobile Base 4. Fortunately the radio transmissions were very clear and he was able to do this as we sat munching our hard-tack biscuits. Little did we know that this was to be a turning point in our own journey: it was by now 3 o'clock and my own vehicle wouldn't start. Kriss delved into his rucksack and found the appropriate tool. He lifted the bonnet of the ailing machine and hit the carburettor hard. The engine burst into life immediately. Kriss's mechanical ability was immediately acknowledged and a precedent was set for the rest of the trip.

For the next 10 kilometres my vehicle limped along and at times the engine cut out just as we were climbing a steep ridge. As we lost momentum we dug in deeply, the engine stalled and wouldn't start. This time the magic spanner did not work and the problem compounded as the battery became weaker and we continually tried to start the engine. Suddenly the engine started again and with a tow from Kriss we got out of one hole only to get stuck in another. Then it stalled again and nothing would coax it into life. Clearly, there was a fairly serious problem with the carburettor. As we cleared the sand from the fuel filters two Bedu appeared with their camels. These were the first we had seen despite passing a number of *barusti* shelters. We offered them some tea from Kriss's flask and Bob asked if he could ride their camel. With one leap he lay sprawled on the back of a very surprised camel which decided to take off at great speed. Hamid was left holding the carburettor and the two Bedu literally fell into the sands as their prize camel tossed Bob over its shoulder into a sand dune. An hour later we were on our way again.

We made good progress for the next 20 kilometres or so and at 5 o'clock found a flat spot to spend the night. We were confident that we would be able to traverse the high dunes the next day, and to show that he was pleased with how things had gone Kriss let off a flare while I was quietly doing my ablutions behind a distant dune. I slept in a hammock strung between two vehicles, pleased to be back in the Sands.

Next morning my vehicle had to be jump started as the battery had short circuited during the night. This was not a good start but we made good progress until we reached the first of the high dunes. It was a towering mass of sand, pure and uncontaminated by vegetation, a mass of different pinks. Alas the vehicle packed up completely. We had no choice but to tow it to the nearest Bedu track and find a route back to Taylorbase. We had crossed a track earlier and headed back in that direction. It was while we were doing this that Kriss's own vehicle started to complain. Towing a dead vehicle through the sand is not easy and often both vehicles became badly bogged. The hopelessness of the situation became apparent as the light faded and both vehicles became well and truly stuck. Kriss informed Taylorbase of our predicament. Extricating his vehicle and leaving Bob, Chris and myself to make supper, Kriss and Hamid set off to try to find a better route out of this part of the Sands.

He headed due west but with his own vehicle becoming less reliable, traversing the dunes in the dark was a hazardous business and he spent over an hour trying to get out of a large hole. Having failed to find the road he returned for supper. I had stood on a nearby dune with a torch pretending to be a lighthouse. So now we were in the middle of the Sands with two useless vehicles. Kriss diagnosed a faulty water pump on his vehicle. As any search party could also get stuck trying to reach us, we decided to have a go at getting Kriss's vehicle back on the road by transferring the water pump. To do this in a garage is quite a feat let alone in the middle of the Sands under failing torch light; it was well past 3.00 a.m. when Kriss finished the job, not before I had fallen asleep dropping the torch into the engine.

The following morning Kriss's vehicle was as good as new. Mine was now in many pieces. We left Bob and Chris with enough food and water for several days whilst we headed back to Taylorbase to get another vehicle and the required spares to repair mine. There was concern that we too would get deeply stuck in the sand, giving any rescue party from Taylorbase a very difficult time. By heading north west this time we crossed many tracks leading to Bedu camps and the number of goats we saw increased. Kriss often stopped to ask the way and we made Taylorbase by nightfall.

Two days later we were back with a new vehicle and continued south, stopping every 10 kilometres to sample our position. We arrived back at the point where we had turned round before. Our tracks could still be seen. We camped the night on the edge of the large dunes. They rose high into a small hill, the crests and ridges forming a jumble of patterns. The wind blew that night and as we ate our supper we talked of the Wahiba bear. We didn't make a fire that night but sat round the light of a 'sand lamp' – made by pouring petrol on to an old can full of sand. I felt a strange spirit that evening – as if we were being watched. The others commented on this as well. It may have been the apprehension of the task tomorrow. We hadn't seen any Bedu camps for some time and the lack of goat or camel droppings indicated there hadn't been any herds this way for some time. It was cold that night and we huddled into our sleeping bags.

I awoke very early, before the sun had risen, still with the feeling that we were being watched. I looked towards the high dunes, surveying the area as a mountaineer might before setting off. Perched on the side of the highest dune was the outline of a dark figure crouching down. Although indistinct from where we were, these were the eyes that had been watching all night. I woke the team to explain we were not alone and jokingly said that this mountain of sand had a special guardian. Maybe this was the bear. We waved but the figure didn't move. We drank our morning coffee in silence trying to come to terms with the day ahead. The figure didn't help.

Of course there is no such thing as the Wahiba bear. But the Sands make individuals humble; they are difficult to come to terms with in the sense of coping with their size. While in a team and on the end of a radio, it is easy to forget their enormity. This tells us something about the Bedu and their nomadic life. They perhaps need that space in the same way that a city dweller needs the hustle and bustle and bright lights.

By 6 o'clock we were packed ready to move on, the vehicles fuelled and all the stores tightly tied down in the back. As we left that site the figure of a tall beautiful Bedu woman quietly slipped behind the distant dune. For the next six hours we had a most exciting time, with much whooping as we roller-coastered through the largest dunes we were ever to encounter. At times it was like a bob-sleigh course and we had to maintain a fast speed to avoid falling into some very deep holes. Views from the top of the largest dune were breathtaking, a sea of sand waves stretching as far as the eye could see.

We must have crossed three such mountains during that day. But as the day progressed the going got easier and we made faster progress in our bid to reach the coast by nightfall. Occasionally we followed the course of the main swales, which were now in a north–south direction and by mid-afternoon we were amongst the white coastal dunes. These are completely different, like rolling farmland covered in snow. Each dune may rise over 100 metres but the slope is long and steady.

We hit the coast near Quhayd before sunset. We had travelled over 120 kilometres through the Empty Quarter and Bob and Chris's transect was complete. I doubt it was a journey that would be repeated for many years; certainly it shouldn't be done without considerable logistic and radio back-up. Driving up the coast and back on to the M3 track we drove fast following the pickets all the way back to Taylorbase in time for a late supper and to tell Beck and everyone about the first sighting of the Wahiba bear.

CHAPTER EIGHT

A Journey with Said Jabber Hilays

Our days were a joy, and our paths through flowers.

Thomas Hardy

'Yalla, yalla let's go,' shouted Said as Dhiyab and I climbed into his truck and we drove off at great speed. Said had warned his family we were coming; we laughed as we fled the hurly burly of Taylorbase for Sayh al Samrah in the western area of the Sands where many Bedu families were camped, including that of Said's sister Latifa.

Said asked where my camel stick was and I had to tell him that it had been mislaid. He reprimanded me for wearing boots and socks, saying I looked more like a 'weekender'. More importantly, he wanted to know why I wasn't wearing a *dishdasha* like Dhiyab. There is only one way to be properly dressed to ride a camel in the desert and that is in a *dishdasha*. I had to admit that we hadn't been able to find another that fitted and Said accused me of eating too many dates. I had to agree, but I didn't want to let him know that I had been a little wary at not being able to pad my behind in this attire. Dhiyab was dressed correctly, however, and was clearly the one who would be the true Bedu; he confirmed he was wearing no further protection under his *dishdasha* and *wizar*.

But then, Dhiyab was always at home in his *dishdasha, masarra* and *khanjar*. I was the odd one out but felt honoured to be able to join the journey. Dhiyab had planned this for several weeks now and even Said was excited at being able to revisit the past with his new companions. He admitted he hadn't ridden a camel for several years because there was no longer any necessity to do so – the pick-up truck was used throughout the Bedu community.

'Wilfred Thesiger warns that the first four days are the worst,' I reminded Dhiyab. I wondered if we would really suffer so badly. The aim of our exploration was to experience the traditional life of those who had made the Sands their home. We were to spend several days travelling by camel through the centre of the western rangelands in order to learn more about the original pace of life and the cornerstones of the Bedu existence. It is possible to appreciate the true size of the Sands once you have reached its remotest corners courtesy of your camel. Water and green grass become the essential ingredients. Desert manners are the root of Arabian hospitality and both Dhiyab and I were to experience a warm welcome many times in the next few days, on more than one occasion when we were on our last legs.

We were travelling light; we could have hired further camels to carry a surplus of supplies but that would not have been the normal thing to do. Dhiyab and I had discussed this at length beforehand. Disappearing into the Sands without much chance of being rescued with only small amounts of food went against all my expedition rules, but if we were to be desert travellers we had to do it properly. Water in two new canvas water bags, dates, coffee and rice were to be the main ingredients of our supplies. Dhiyab and I were only allowed to carry minimal personal kit, the most important item being a modern sleeping bag – to pad the saddle.

Said stopped the truck outside the supermarket in Mintirib to collect some provisions for his family, including some fresh meat. Dhiyab and I remained in the cab. We noticed Said pointing at us and this was followed by some raucous laughing; perhaps we were suspected of following the European romantic view of the desert too far. Having re-read Wilfred Thesiger's account again, I did for a moment wonder if we were being a little too idealistic. However, that doubt did not linger long.

Ten minutes later we turned off the road on to a sandy track leading to the Wadi Batha. Across the wadi was the head of the same swale that Wilfred and his companions had travelled through some 38 years earlier. This swale went due south to Tawi Haryan, one of the wells at which we would be stopping. The swale we were to follow was three or four swales to the west. Soon we were at the Wadi-sand boundary that delineates the true edge of the Sands. Said drew a line in the ground with his camel stick precisely where the darker gravel sediments of the wadi met the soft sands of Wahiba. From here on, he explained, it was the true land of the Bedu. We were in the Sands proper.

We raced on down the swale. Latifa's encampment was some 50 kilometres due south. The track was well marked as it was one of the main routes to the western grazing areas and would have been in constant use during the last three weeks or so. The dunes either side of the swale were some 100 metres high and the swale was about a kilometre wide, its floor firm and covered by a variety of plant life, particularly the now green *Panicum turgidum*. Said was pleased with the obvious growth of the vegetation. He had three of his camels in the area and wanted to locate them during our ride. We had flown over this area in January; it was uninhabited then and we had seen no evidence of either Bedu occupancy or goats and camels. Now that the area had flourished after the rains there were a large number of camps and many Bedu had brought their herds from the north and south. Nomads cope with such change both in the short and long term; their versatility is an enviable trait, particularly useful if the processes of change are rapid.

Said drove far too fast. He started to sing and he always drives even faster when he sings. Dhiyab joined him and they made up verses about the ride and what would happen to us all. As they both sang the truck careered along and I hung on tight to a handle on the dashboard. There would be no time to stop if a truck coming the other way were also being driven by a singing Bedu. I wondered if our insurance cover included camels. Dhiyab and Said were now in full swing, occasionally roaring with laughter. I am no singer but joined in all the same, the sand tyres on the firm ground giving us a very soft ride.

We arrived at Latifa's *barusti* shelter at 4.30 p.m. but she was not there. We parked the truck a little way off and unloaded. Two camels brayed our arrival and we were welcomed by her daughter, who asked us into the encampment. Made from date palm leaves and stalks, a small amount of corrugated iron, canvas tenting and chain-link fencing, these shacks are totally practical. They bear little resemblance to the black tents that are known elsewhere in Arabia. The palm gate swung open and we squeezed through to sit down on a brightly coloured rug and make ourselves at home. The rugs that formed the side of the house had been rolled up and a cool breeze was passing through. Said started catching up with the news while Dhiyab and I tucked into a large pile of dried dates; they were delicious and were to be the first of many we were to eat over the next four days, for as soon as you arrive anywhere in the desert you are offered dates and coffee. Today we were also given a butter-based sauce to dip our dates into which Dhiyab sampled with relish, although I thought it an acquired taste. One of the young camels poked her head over the side of the fence to remind Latifa's daughter that it was past feeding time. She gave the camel a handful of dates from the bin and patted her affectionately. The camel munched through the dates with as much noise and saliva as a camel mouth can muster; camels love dates.

Said departed to check that our camels would come later that night. It would have been impolite to remain in the hut while he was away so we finished our coffee and went back into the open and up the small dune behind the camp, where we sat down in the sand.

Before us the hubbub of a small Bedu encampment unfolded. Unusually for a Bedu woman, Latifa had her own pick-up and she returned with a great pile of firewood, harvested from some nearby *Calligonum*. Goats appeared on the dune opposite and headed towards the house, apparently without any shepherd. Latifa's daughter was tending the fire and had begun cooking the evening meal – she had welcomed Said's meat as she was having to cook for a larger gathering than usual. Chickens were pecking around searching for pickings from the kitchen. The young camel was now getting fractious. Camels are rarely tethered – they are hobbled instead. A short length of nylon rope is looped and twisted to form a figure of eight and the front hooves go into the holes, leaving the camel only able to take steps of just a few inches. It is a satisfactory way of preventing the camels from wandering off too far.

Said returned with the news that the camels would be male. We were pleased to learn this as male camels are more docile and dependable and therefore infinitely preferable for everyday use – and for beginners. Females tend to be the racers, being smarter and faster and, since they are highly strung, very much more demanding. We would have some 'driver training' as soon as they arrived. Said loved to take the mickey out of the project whenever he could and he was unable to resist reminding us of those early days when we were all getting to grips with Land Rover driving in soft sand. He explained that the camels also had five gears and that they were the latest models that didn't need double declutching – although he wasn't sure if they had handbrakes or not. His eyes flashed mischievously as he observed our apprehension.

Said said we must get to bed early as we wanted to leave first thing the next day, but a Bedu's idea of a long night's sleep is different from ours. Discussions go on until long after midnight and the fire is stirred a few minutes prior to the first light of dawn, which, at that time of year, was around 5.20 a.m.

Soon a number of visitors started to arrive. The first group came from over the hill, walking with two magnificent camels in tow. These were ours and they were enormous. Shortly afterwards a Bedu in a small blue truck arrived with a camel tethered in the vehicle.

Said then introduced us to the team – Furayhin, Khumaysan and Sawghan. And what a trio they were. Said's camel, Sawghan, looked as if he might live up to his name which meant 'good natured'. If I had been given the choice he was the one I would have opted for. Furayhin was to be Dhiyab's. He was the most thoroughbred in appearance, a magnificent-looking beast from the Farhah breed for which the Al Wahibah are particularly well known. Khumaysan, from another famous breed, the

Khumays, was mine. He didn't look quite so aristocratic – indeed, there was something a little roguish about him. I went to make my acquaintance. He was taller than I expected and I could only just see over his neck. He was no beauty and did have a distinctive smell – the scent glands on the back of his neck were obviously highly active. He was blowing bubbles and thick white saliva was spluttering all around and dribbling down the side of his mouth. For a moment I joined the ranks of the camel-wary.

Said prepared his camel for driver training. It looked easy – you jump on the back and just ride as you would a horse, but without stirrups. He asked who wanted to go first and, wishing to show my keenness and to hide my apprehension, I sprang to my feet. Sawghan was lying down on his haunches with his front legs tucked under. Now camels are trained to rise as soon as you are on their backs and they do so in three movements which when done fast, catapult you up and over. To contend with this you need to anticipate and co-ordinate your own movements with your beast. Said held Sawghan down, whispering commands into his ear, but the camel sensed my fear and was restless. I took hold of the little wooden saddle in both hands and wedged my fingers under the strap that went under the camel's belly before leaping aboard. Before I knew it I had been thrown forward so fast that my nose became buried into the camel's mane and my posterior pointed upwards. I landed roughly in the right place upon the padded saddle that Said had carefully prepared; it was surprisingly comfortable considering that camels have very pointed and bony spinal columns. Said took the rein and we walked up and down the swale for a while. A camel's gait cannot be described – it defies any known rhythm. Like the camel itself, it might well have been designed by a committee.

The ride had been a success. I dismounted and Dhiyab took over with real confidence. Said was pleased; he gave us our driving licences, as it were, and remarked that tomorrow would be 'maa fii mushkilah' (no problem). The camels were hobbled in the lying position and given some alfalfa to chew. They were content and showed it by chewing their supper loudly. Camels have teeth which give a variety of notes as they masticate and these musical teeth will sometimes rehearse a monotonous concert throughout the night.

We stockpiled our rations and equipment ready for the early start and made ourselves comfortable around the fire. Said started to brew up some coffee. Once the sun has set the desert gets cold remarkably quickly and both Dhiyab and I wore all we possessed which included, in Dhiyab's case, a pair of locally made goat-hair socks. The group grew in size as the evening wore on and at each new arrival we all rose to exchange traditional greetings; Dhiyab and I noticed expressions of amusement as they surveyed their eccentric guests. Latifa and her family were still cooking what appeared to be a feast of gargantuan proportions. The owner of the blue truck and Khumaysan who had, until now, stayed in his vehicle crawled out of the cab and on to the sand. It immediately became clear why he had not previously alighted – both his legs were stunted, very thin and too weak to bear his weight. He crawled over to join the throng around the fire. He was a cheerful man who joked about Khumaysan's fitness – he had been disabled since birth but this had not prevented him from being able to ride a camel successfully.

Latifa brought us a huge tray of rice and a large bowl of meat and gravy, which she put down on the sand nearby. As is the custom, she then returned to her house while those who were staying for supper crouched round the tray together. The feast was large enough to feed 20 at least and we all attacked it with gusto, Said joking that we should eat as much as we could as we might not find a good hotel tomorrow, by which he meant a Bedu encampment that had sufficient rations to lay on such a meal. One of the advantages of eating in the dark is that you don't differentiate between flesh and guts, tongue, brains and other bits – and, very occasionally, the prized eyes. However, since we knew this was the meat we had brought from town we gorged all the more. For us, the limiting factor was how long we could perch on our haunches with one leg tucked under the other. Being somewhat top-heavy, I tended to get pins and needles within ten minutes or so. This meant I would have to stretch painfully and start again.

After the feast Said, Dhiyab and I returned to the fire, having bid farewell to our companions. They had wished us well on our expedition and said they might return to watch the launching tomorrow. We suggested that they give it a miss this time, but we were gaining confidence now. We were already beginning to feel at home and, more importantly, that Said wished us to feel that we belonged to his way of life, his family and his Sands.

It was just after 5.00 a.m. when I heard Said clinking the metal pestle and mortar as he ground the coffee and cardamom seeds. He had learnt it made an excellent alarm call and this morning he was clattering it with extra gusto. Dhiyab and I stirred. It was cloudy, which meant we wouldn't see the sun until mid-morning. My sleeping bag was wet, but not unduly so. Latifa arrived with a full Arabian breakfast of warm camel milk, a wafer-thin kind of pancake bread, oranges and coffee, all of which we found delicious. Said was anxious to be going and he helped us load in turn. The way you saddle your camel is the most crucial thing you do in the day – get it wrong and you are in discomfort from the moment you mount. The saddle horn is relatively small; the saddlebag is woven from goat hair and is traditionally in a red and brown pattern. It has two compartments, which hang either side of the sharp spinal column. Into each go your items, the bulky ones at the front so you can dangle your legs over them, the soft ones, such as sleeping bag and woolly jumper, arranged to form a landing pad. Dhiyab mounted and started off down the track southwards, riding as if he had been doing it for the past 20 years. Only his pale skin gave away his true origins.

My mounting was both spectacular and memorable and, unfortunately, half the Bedu camping near Latifa had turned up to watch us depart. A row of smiling faces was the last impression I had before Khumaysan threw me over his shoulders and back on to the sand. My fingers, which had gripped the saddle strap like limpets, had been wrenched so hard that I heard a little ping at the end of my middle finger. I could hear the roars of applause on the bank above. Said immediately asked if I was '*takliif*' (in pain). I replied that I wasn't, picked myself up from the sand and brushed myself down. Said made the 'kher, kher' noises necessary to get Khumaysan back on the ground and I made a second attempt, this time successfully. My success, however, was short lived. Khumaysan was a racing camel, and racing camels are spurred on by cheering. The roars of appreciation from the launching party and my own cries stirred him into an immediate gallop.

This was no way to learn to ride a camel. We were soon in fifth gear and Said had not told me where the brakes were. I tugged on the rein but camels are controlled by voice and camel stick, not by the reins. This was perfect entertainment for Latifa and friends as they watched Khumaysan take off down the swale after Dhiyab. My grip was loosening fast and I was beginning to bellow in panic. I ran through my limited vocabulary, urgently trying to find the word for STOP in Arabic. Khumaysan and I were not communicating at all well by now.

Before my eyes flashed an experience that had happened to me when I was six years old. I had been put on to a very large, untamed Kenyan horse, which had decided to gallop directly from a walk. I can still remember the fear of trying to hold on with my toes and fingernails as I slowly slid off round the horse's neck, hitting the hard African ground like a sack of potatoes. This was an Arabian *déjà vu*. My bellowing, rather than curtailing Khumaysan's racing instincts, simply stirred forgotten energies. I had by now let go of the reins and wedged my fingers under the saddle straps while my feet and toes gripped his sides like claws, but to no avail. Slowly but surely I started to fall forward round the camel's neck, which was even less designed for holding on to than a horse's. I hit the sand hard and somersaulted. I could hear more roars in the background as I lay crumpled, winded and shocked. It was some minutes before I realized that there was concern on Said's face as he helped me up.

'Are you in pain?' he enquired again. I couldn't reply. My pride was too hurt.

Khumaysan took some retrieving and indeed it was his owner who caught him by going off in the blue truck with a helper. I mounted for the third time and wedged my now swollen fingers back under the saddle strap. Said mounted his camel and took my rein to make sure no further foul play could take place. I turned to wave at Latifa and friends, sure that they were convinced that Said had bitten off more than he could chew with these two apprentices. I shouted we would be back in four days – 'Inshalah' (God willing).

Dhiyab had travelled a long way ahead and had missed the shenanigans. I knew Said and he would have a good laugh about the learner driver and no doubt there would be a song. Dhiyab looked confident and comfortable. He remarked later that I looked a little pale but didn't notice my bleeding fingers under the camel strap.

Now it was just the three of us again, heading due south into the heart of the Sands. We were following a swale and, on the horizon, a cross dune formed a wall of sand we would have to surmount. In the distance there were the remnants of some night mists. My inauspicious start had been unsettling, but the journey itself soon took command and gave me the most enjoyable moments I had during my days in Oman.

Said asked if we wanted to trot and we both paled at the suggestion. He burst out laughing and told us that we had over 50 kilometres to do today if we were to reach water or other Bedu encampments. By now it was just 7.00 a.m. and we had settled into a firm walking pace, with Said still holding the rein of Khumaysan. We both knew that we would have to trot soon and so with reluctance we agreed to have a try.

Trotting without stirrups means you are thrown into the air with every step and land back with a bump. Timing is critical. However, to Said's amazement we managed it with little fuss and he burst into song, waving his camel stick to the rhythm.

The quiet of the desert was interrupted only by the sloshing of the water in the canvas bags, the grinding of the camels' teeth, the plodding of their feet on the sand and the tapping of our sticks on their shoulders. Otherwise there was complete and utter silence. Such silence is rare wherever you are; it becomes all-encompassing and I have to admit I dozed off into a trance, mesmerized by the rocking gait.

At about 9 o'clock I noticed the sun beginning to break through the clouds overhead. The freshness of the night mists had only just evaporated and before long the sand would heat up fast. Both Dhiyab and I covered our heads with our respective headdresses. Our pace was steady and we were getting into our stride with the occasional trotting session. We estimated our speed to be an average of 8 kilometres an hour – not bad for novices but would we reach the grazing grounds today?

The landscape was flattening out slightly, with the dunes more evenly covered with a fine veneer of new, green, woody vegetation. We had left the first swale after traversing the wall of sand that crossed it and had dipped eastwards into the neighbouring swale. We disturbed two gazelle, which ran on ahead of us. The area was certainly becoming very green indeed, with the *Panicum* and the grasses sprouting vigorously. Small yellow and white flowering plants provided splashes of colour. It was incredible how much more we saw from the back of a camel and I remembered Wilfred Thesiger's distaste for travelling in vehicles. Lizards, skinks, large dung beetles and busy caterpillars carried on their business right under the camels' feet. Said pointed out the day's activities that had already left their imprints in the firm sand.

'Look, there are wildcat tracks . . . she was just scouting around; here five gazelle passed not so long ago and two of them were young ones; there are fox tracks and, look here, a small mammal had a squabble with a bird of prey.'

Said was in his element as a guide and I could imagine that one day travellers might be ticking animals off in their handbooks as they do in the African national parks. For a moment I felt that all members of the team should have the opportunity to travel the Sands in this way. We were lucky today, because the sand was relatively firm from the rains and also because the overcast skies had delayed the burning heat of the day. I asked Said if we would meet any other Bedu travellers and he assured us that we were the only ones on camels this weekend. I wasn't sure whether to believe him or not.

In one swale we came across some camels and Said took off to identify three animals that were possibly his. We followed behind as he cut across steep dunes. My grip tightened under the saddle but I was now using the camel stick more professionally and was even striking up a degree of communication with my beast. Said was clearly disappointed when none of the strays turned out to be his but he did know who they belonged to from the brand marks and would be able to report the news when we got further south.

Without doubt, Said is immensely proud of his camels. The relationship between camels and their Bedu owners is legendary. These animals may be odd in appearance but they are a perfect companion for the desert – imagine the days before trucks, established wells, roads, signposts or other litter; your vehicle had to provide reliable transport which could go for long journeys with very little refuelling, and be able to cross soft, hot sand. At night the camels gave shelter against the storms, nourishing milk to drink and companionship against the *jinns* of the desert. It is said that there are over a hundred names for the camel, or *jamal* as they are known in Arabic, and their name is derived from the same root word as 'beautiful'. The Bedu have strong sentiments towards all their camels and Said was no exception. My own friendship with Khumaysan was growing and I think we were coming to terms with one another. I sometimes missed the sound of his teeth grinding. The bubbling soft palate roar I could still do without, particularly when flecks of white foam were thrown my way.

Dhiyab explained that the Omani camels have always been considered to be among the finest in the world, particularly when it comes to camel racing, which must be the national sport of Oman. I recalled the meets at Seeb and Dariz, where we had seen some of the best animals in the land. Prices for these were rocketing, especially with buyers from the United Arab Emirates offering high prices for proven winners.

Dhiyab and Said started singing again. They took it in turns, Said offering a ditty in Arabic to which Dhiyab replied in traditional Scottish verse. I remembered the first time I had heard Said singing – his tunes hadn't altered much but he did now have a greater repertoire of verse, much of it about the project, its members and the times we had spent together in the Sands. As for Dhiyab, I had first heard him sing at a folk club in Exeter, where he had received loud applause from the regulars. Folksongs worldwide have passed on information about events for generations and Dhiyab's song, had Said been able to understand it, would have given him an insight into the cultural habits and festivities of those living within the borders of Scotland in the eighteenth century. Said's ballad would later be reducing his Bedu companions to bellows of laughter as he relived his time in the Sands with us. Here was desert data being communicated in verse.

It was coming up to 11 o'clock. Our backs were now aching considerably and Said could see we were in pain, although we both denied it, rather too quickly. He knew we were lying and comforted us by saying we would stop soon for coffee and a large lunch, but he had been saying the same thing for some time. It was certainly hot now; the clouds had disappeared and the sky was clear from horizon to horizon. We had made good progress, although diversions to look at camels and pauses to pick up dropped camel sticks had delayed us slightly and we were still a long way off from the grazing grounds. Eventually Said caught sight of a large patch of *Calligonum*, the only sizeable clump for miles around, where we could base ourselves for a long lunch break. Dhiyab and I were immensely relieved; he had only backache, but I could feel blisters forming and my fingers, which were still clenched round the saddle straps, were now numb and caked in blood. We had been riding nonstop for nearly five hours.

We dismounted and I fell into a heap. Dhiyab was more agile and helped hobble the camels, including Khumaysan, then began collecting wood. Said immediately started to tie the *Calligonum* bush together and propped part of it up to form a cover, over which he put his blanket. This provided an area of about four square metres of shade, enough for three to sleep. Just as I thought things were improving, Said suddenly gave a cry of anguish; our new canvas water bags, which had been providing that comforting sloshing noise, had leaked very badly. This meant that we could only have dates and coffee for lunch, since we did not have enough water to boil rice. The sun was now directly overhead and Said, Dhiyab and I, having drunk sparingly from the half-empty water containers, scraped the sand to form a shallow hollow and settled down under the bush together with our headdresses wrapped round our faces and heads to keep out the sun and the many insects that shared the bush with us. As far as I was concerned we were in the middle of nowhere and I had never been so grateful for shade in all my life.

By 3 o'clock we were back on our camels, Said leading at a steady pace. He knew we had a good three hours' travelling before it got dark and if we wanted to have a good meal tonight, and water the camels, we would have to maintain a good speed. We trotted for most of the way and I decided that as long as we didn't start cantering I would be able to stay on. The coffee we had had for lunch was the most refreshing drink I had ever tasted, with cardamom seeds giving it a distinctive flavour, and for the next three hours I dreamed of the subsequent cup which I knew we would get as soon as we arrived at the Bedu encampment. I was in a worse state than my companions as I really wasn't acclimatized to the midday sun. I looked across at Dhiyab but he was absorbed in steering his camel and keeping his balance. Said had turned down our request to walk for a while on the grounds that if he didn't get some supper that night he wouldn't be able to sleep.

We were heading in a slightly south-westerly direction, remaining in the same swale but having to cross a number of smaller linear dunes that ran diagonally across it. Some of these were remarkably steep and had I not been in a trance, with one hand stuck firmly under the saddle, I would have marvelled at Khumaysan's sure-footedness and his ability to cross the most yielding sand; his wide, padded feet were like soft tyres which spread neatly with each step.

We would have been happy to call it a day by 4 o'clock but Said did show genuine concern for the first time. We hadn't seen any sign of camp or truck all day and I think he was a little worried about our capacity to manage without water. It is very bad manners to arrive at a camp where you expect to receive hospitality too late in the evening as this may inconvenience your host. It is proper to arrive well before dusk to give the hosts plenty of time to organize a meal while it is still light enough to see. We were beginning to marvel at Said's navigational confidence; he was following a direct route to a known spot on the horizon.

The next hour continued in a desert fantasy for me. The hind legs of Said's camel, directly in front of me, reminded me of the elegant posterior of a slim lady wearing brown tights, leotard and high-heeled shoes. There was definitely a similarity and I wondered if Dhiyab had also noticed. Khumaysan was beginning to sweat now and the glands on the back of his neck were pouring a thick treacle of musk. I felt in awe of those who managed to travel under such conditions in the summer when the temperatures would be several degrees higher and dry winds would carry biting sand.

Dhiyab gave a cry as he recognized a range of dunes that he had visited earlier in the project; he knew now we would be with Bedu before nightfall. Said asked us to quicken our pace and even the camels sensed we were nearing our destination. The sun was beginning to go down as we caught sight of a herd of goats and sheep. As we rode over the next dune the views to the south, east, west and north were quite magnificent and I wished I had the energy to get my camera out, for here was evidence of the green desert and there were prosopis trees in the distance. It was nearly 6 o'clock when we dismounted and, although it was late, we were welcomed into the single *barusti* camp of one of Said's Bedu companions.

Said's friend, who was also called Said, asked us to sit on the rug that he laid out on the sands and fetched dates and some hot coffee in a large, brightly coloured thermos flask. This strict code of hospitality has saved lives in the past and it was the welcome Dhiyab and I and, I am sure, Said too, had been dreaming of ever since we had learnt that our water bags had leaked.

Said and Said Two had much to catch up on as they hadn't seen each other for some time and had many mutual friends. Said Two was an old man who lived alone, which was unusual. He had heard of 'the wolf' and indeed Dhiyab had met him twice before during his travels and was able to recognize him, even though he had met so many Bedu during his visits to camps and it is not easy to put names to faces only seen by firelight.

It was not every day that Said Two had visitors by camel and it was an excuse for a feast. Said Jabber insisted we shouldn't have anything special. Our host insisted on killing a goat; Said was adamant that he should not. A torrent of dialogue ensued. Dhiyab and I left them to their discussions and went to sort out our camp items so as to be ready to climb into our sleeping bags immediately after supper. We were both dog tired, with raw behinds and very sore backs. After eight hours of riding on our first day we would surely wake to stiffness in the morning.

Said Two won the argument and we did have a feast, tucking into goat and rice like vultures. The two Saids discussed all important desert issues; where the new grazing lands were, what prices camels were fetching, what the rumours were concerning the scientists looking for oil and so on. Said Two was head of one of the subdivisions of the Al Wahibah tribe and Said Jabber was head of another, so they did a great deal of talking on tribal matters and issues of local politics. Dhiyab and I were by then tucked up snug in the hollows we had scraped in the ground, our feet facing into the wind to stop our bags filling with sand.

The wind was still blowing early next morning. Said Jabber shook me gently and explained it was time to get up and that coffee was being served. My face was covered in sand and my lips were like sandpaper. I was stiff, so much so that I did not want to move. The wind had kept the night mists away but my bones could still feel the moisture in the air. I opened my eyes and could only just see the first traces of light. It was that short period that is neither night nor day and I wanted just a few moments more in my sleeping bag, but I could hear Dhiyab and Said laughing about their aches so I forced myself to rise and join them. I was becoming addicted to the coffee and had more than three cups before I felt ready to face the day.

Now that we had reached the grazing grounds Said wanted to head a further 15 kilometres south to the area known as al Khawasiyat, which is bound by the prosopis woodlands, since he was keener than ever to find his camels. He now expected me to be able to saddle my own camel and Khumaysan and I were left together to sort ourselves out. To get a camel to lie down you have to make a kind of rasping noise in the throat in short, staccato bursts.

'Kher, kher,' I asked Khumaysan, tugging at the rein, camel stick in hand.

We were getting along much better now and he sank on to his haunches and knelt down. Once down, the hobbling cord is tied round the tucked-in hoof and the thigh to prevent any unplanned rising. To do this you have to crouch right down under the camel's head. I could feel Khumaysan's hot, sticky breath at the back of my neck and for a moment the hairs there stood on end, but he didn't show any sign of disrespect. I was beginning to understand Said's affection for the beasts.

Said was watching from afar and swung his stick against the sand to show his pleasure. I draped the camel bag over Khumaysan's back and fastened it to the small wooden tripod that remains tied to the camel's back and serves as an anchor. The straps that attach this firmly to the camel were the ones that I had wedged my fingers under the day before and I could still see traces of blood. I placed a second goat-hair bag over the back and secured it in place. A blanket, my sleeping bag, and my own small sack of items were all tied to form a seat. There is a great art to this and Said was letting me learn it the hard way.

We mounted and after the initial buck I settled into the saddle I had just prepared. It was surprisingly comfortable for my blistered backside, but I was not sure I had the tenacity to cope with a further eight hours' ride today. We said goodbye to Said Two and left the security of his little camp. Said Jabber asked if we were comfortable soon after we had slithered down the first dune and we both asserted that we had never felt better. The views that morning further astounded us. We were crossing a particularly high ridge and again the swales to the south and north and the 50-metre ridges to the east and west were very green indeed. Dhiyab began to sing some West Country ballad and Said joined in with his own song. This was very different to any area of the Sands I had visited and would have been a dull grey landscape prior to the rains, but now it was difficult to believe that the substrate was still all sand.

We travelled at a fast, steady pace for an hour or so. Two encampments appeared on the horizon and we headed for the one directly in our path. A shout beckoned us as we drew near and we welcomed the opportunity to have an early morning break. We dismounted from the camels and exchanged greetings with the Bedu – a man called Mabruk and his family, who had recently arrived in the area with their herds. They were pleased with the pastures and planned to stay for about four weeks or so. The children, who were playing on the hillside opposite, laughed and joked when they saw the odd-looking visitor with Said and his Bedu companion. The wife tended to the fire and started to warm some goat's milk while we settled on a rug and again tucked into dates and butter sauce. Such informal entertainment can soon turn into a full-scale meal and Said was keeping an eye on her to make sure she didn't start preparing lunch.

Said and Mabruk began to talk earnestly. Mabruk wanted us to stay for lunch and was very insistent, even though it was still early morning. Said was equally insistent that we wouldn't, eventually giving the excuse that he wanted to look for his lost camels and would have to keep moving. Mabruk finally accepted a compromise – he would go ahead in his truck to Ali bin Hamad Lahush's camp and tell them we were on our way to have lunch there. Then, after lunch, they would lend Said a vehicle so that he could drive about looking for his camels while we rested up in the camp.

Suddenly the time constraints of this journey, dictated by the need to be back at Taylorbase on the promised day, gave our desert expedition a different pace from what might have been normal. I felt an intruder who had no right to be there. Said was pushing the pace of the journey to suit us and this meant we had to disentangle ourselves from the warm hospitality being offered. The project as a whole suddenly seemed an intrusion into the life of these masters of the desert environment.

I began to understand the reason why the Bedu will remain in the Sands, even though they have the wherewithal and opportunities to settle in the margins. The Sands are their home and there is no reason why they shouldn't remain. The pastures offer excellent grazing and their own management systems which prevent overgrazing seemed to be working well. Now that the pick-up could carry and deposit water tanks anywhere in the Sands it was no longer necessary to be dependent on wells, although they were still valuable. Dhiyab spent many hours on our trip discussing the changes, interested to learn more about the pressures that oil-wealth had brought and whether further development would be influencing the next generation. Matters of health, education and status are not subjects easily discussed with foreigners and Dhiyab had worked hard over the previous three months to secure the Bedu's trust, friendship and respect. I was the outsider and I hoped I could remain unobtrusive, for I wanted to see their life as it really was.

We travelled on in silence, heading south west towards the prosopis woodlands, which were now some 10 kilometres away and becoming much more visible. How the day would unfold was of no concern at that moment. The Sands were a large, forbidding and lonely arena and I focused my energies on ignoring a growing thirst, a rising sun and pains everywhere. However, the desert is always full of surprises and it was to turn up a trump card. A most wonderful sight made me forget my aches for a short while. Over a ridge there was a small swale covered in a carpet of little yellow flowers and a profusion of other plant life that had recently responded to the rains. There were a few camels grazing, two of them with their udders protected to prevent excessive suckling by their young. Butterflies were dancing around us and I had to pinch myself to believe that I was still in the Sands. Said started to sing again.

He also chose that moment to canter and our camels followed suit. This might have been because Mabruk had jokingly said that he had thought we were old men when he had seen us in the distance. For both Said and Dhiyab this represented a challenge. I let out a yell that Khumaysan had heard before. My depression and my delight in the flowers both disappeared as I quickly tucked my camel stick under the saddle and hung on with both hands. This was a survival situation and I was not going to take a tumble again. Dhiyab, head down, was mastering the art quickly and I was keen to do the same. We must have been a marvellous sight as we careered over the next dune and straight into an open plain where a large number of Bedu had set up their camps. It was an exhilarating experience and for a short moment we were an integral part of desert life. The camels had smelt water and were homing in on an area they knew well; we were no longer visitors but members of a group arriving back into known territory.

Ali Hamad's family were close friends of Said's and insisted we should have a long break there – which in fact turned out to be an overnight stop. We were made to feel particularly welcome as it was a family that Dhiyab had visited on his earlier travels by vehicle and they were most impressed he should return by camel. There were a number of shacks, water tanks, camels and chickens, plus a cow, and the place had the air of a well-stocked farmyard. Dhiyab and I kept very quiet when the discussion focused on whether we would be moving off or not. While we wanted to help Said achieve his objectives, we both realized that our battered bodies and blisters would welcome a break. However, lunch was next on the agenda. A wonderful feast unfolded before us with the whole family sitting in a circle watching, laughing and joining in. The children, prompted by Said, started to mime me falling off a camel. This most memorable occasion was the first time I had shared a group discussion with six or seven Bedu women, all with their masks and dark green clothing. One was breast-feeding a newborn baby, while another spun some goat-hair wool. There was much talk about the project and what it was going to do for the Al Wahibah people.

The large lunch and the hot sun helped us sleep. We accepted the bedroom of a son who had recently married, although I am not sure what his new wife thought of such an intrusion. The shelters, like most in this part of the Sands, were made from canvas tarpaulins or blankets over a frame of prosopis wood and stripped palm branches. The sides could be rolled up to let any breeze pass through. The 'tent' itself was some 5 × 10 metres in size, three-quarters of it was carpeted by blankets and rugs. A special collapsible doorway made from date palm fronds was meant to keep out inquisitive goats. Dhiyab and I fell into deep slumbers straight away, as did the camp as a whole. I woke once to see the cow raiding the kitchen area but otherwise didn't stir for some four hours until Said came back from his quest with the vehicle laden with dead prosopis wood, and the news he had successfully found his three camels which were healthy and in good shape.

For the rest of the day we fitted into Bedu life. For a while we joined in discussions, accompanied by more coffee and dates. This encampment had a large number of passing travellers as the old father was a well-known and respected figure. We were introduced to everyone individually and many were pleased to meet Dhiyab again. Said acted as spokesman for the project as a whole, since it was a good opportunity to share our objectives and explain to whom our information would be going. We definitely found an affinity when we touched on the delicate resources of the Sands as these are the cornerstone of Bedu existence and the respect we gave this balance was understood by all those we met. Many people were particularly interested in Paul Munton's aerial headcounts and confirmed that our figures relating to gazelle would be about right.

At about 5 o'clock we were kindly driven by Ali Hamad to a nearby water tank to fill our water bags for the return journey. This was some 5 kilometres away, a journey which took us right into the prosopis woodland that extended in a long, wide arc down the west side of the Sands.

We headed for a particularly high set of dunes which were in the midst of the trees and our host explained that these dunes had been a meeting place for tribal groups to band together before raiding other families. The area has seen many skirmishes, although not for the past decade or so and the sands have covered any evidence that might have remained. The road was well marked and we soon came across a white water tank where we could help ourselves. This was one of the many that were refilled for the Bedu by the Ministry of Water with excellent, sweet drinking water trucked in from Wadi Matam to the north west. The local wells vary from bad to appalling. 'Khawasiyat' means 'brackish well' and the water from these is only used for livestock or some new and very small gardens. The feeling generally was that all local wells were being overdrawn and the watertable was dropping, so the rains had been a blessing as it seemed as if the pressure might have eased.

Our canvas bags were now watertight – it had taken time for the canvas threads to swell, thus forming an impermeable surface. They were shaped like a goatskin sewn up at the legs and neck, the latter being used as the pouring hole closed with a bit of nylon cord. We did see the original versions, which were made from real goats, and they were exactly the same as the *guerba* used in the Sahara.

That night, back at camp, it was the equivalent of Saturday night out and a number of guests had been invited from neighbouring encampments. They started arriving soon after dark, the bright orange headlamps homing into the camp from all directions. As there were so many to be fed, the men of the families had slaughtered two goats and were busy preparing and cooking an enormous meal. A rug was laid out on the sands and we sat round in a small circle on our haunches; it was cold and Said, thoughtful as ever, drew a blanket over my shoulders.

When two or more Bedu are gathered together there is always interminable discussion of camels – lost camels, found camels, what camels have been seen today, what condition they are in, whose they are, where the best grazing is, where they are watering and so on. Nowadays there are the racing camels to provide further fuel for conversation – who has won recently, who is entering for which race, how the training is going, who the riders might be, who had won a prize, new births, new sales, new purchases ... The chatter under the stars, as we sat on the cooling sand in the dark, must have been the way in which information and knowledge had passed down for generations. This format for collective discussion was timeless. The vastness of the Sands added to the cohesion of the group and the comfort drawn from it secured the permanence of desert hospitality.

Some members of the group remembered Wilfred Thesiger when he had passed through in the late 1940s as he travelled from Wadi Althain to

Wadi Batha via the well called Tawi Haryan, one of the major wells of the region which had been used for generations. It was some 4 to 5 hours from where we were now in a north-easterly direction and we would have to cross 5 or 6 swales before we got there.

We left early the next morning, walking slowly for the first hour or so, although we were feeling much better after the long rest and greatly refreshed. I asked Said how he navigated because we were cutting diagonally across the dunes. To me one swale looked like another. Said simply replied that this was the way and promised to get us there directly. He admitted he had not done this journey for some time. We mounted the camels and rode carefully over some of the steeper crests. Our beasts were sure-footed and gentle, but riding down the steeper banks Dhiyab and I had to be extremely careful not to be thrown over the camel's head. Sometimes it was so steep I dismounted and walked the camel down the slipfaces. After going some three hours, the camels' pace quickened and Said assured us that Tawi Haryan would be over the next swale. Sure enough, as we peered over, there were two enormous prosopis trees and the well just to the side. It is unlikely that it has changed since Thesiger's time.

While we were washing, cooking and resting under the larger of the two trees in preparation for our return to Latifa's camp, an old friend of Said's roared up in his brand new pick-up. Jokingly he said he had not seen a sight like this for nearly ten years. Do you want a lift? You must be crazy! Are you sore? he asked. Both Dhiyab and I categorically denied being in any pain and said we wanted to complete our journey by camel in the proper fashion. The contrast between the old and the new could not have been more stark. The Sands themselves had not changed. The nomadic lifestyle had not changed. The innate knowledge had not changed. The pace, only when it suited the Bedu, had. Thankfully they could still choose.

CHAPTER NINE

Seeking Solutions

Desert areas of the world do not get the attention they deserve in the overall environmental arena and thus, although home for ten times the population of the British Isles, do not yet receive sufficient consideration from the world's development agencies. Consequently desertification, which directly challenges the lives of 80 million people globally, is still low on the political agenda, partly because those who live in those areas do not have much political clout, either nationally or internationally, and partly because our real knowledge of desertification is slim.

Mindful of our responsibilities we began, as the end of the project began to draw near, to discuss how our research work was going to be of value to the Wahiba Bedu, to Oman and to other sand deserts.

We were often asked by the Bedu what influence our results would have and, if there were to be changes, when they might happen. Said Jabber was particularly concerned. On one occasion, while we were sitting in his *majlis*, Said began to ask pertinent questions about the future. He and his family wanted to know if our activities might alter life in the Wahiba Sands, whether more scientists would follow and whether tourists would be permitted to visit the Sands. They were important questions that needed an explanation. As a result of our efforts, did we really have a better understanding of the complex processes that make up this sand sea ecosystem? How could our discoveries contribute to the better management of the world's arid regions – nearly one-third of the earth's land surface?

By the end of the project we had come to terms with the Wahiba sand sea. After four months we had worked in most areas, following the network of Bedu tracks that cross the dunes. The pace had not slackened; our combined mileage equalled twenty times round the world. There had been some near misses, but these had usually been on the fast main road to Muscat at night. By and large the dunes were our friends and they did not now seem so frightening. Only once, when one of the Land Rovers rolled on to its side, did they threaten life; but after a radio SOS, the use of flares and a speedy search-and-rescue, all was well. For most, there was now a deep respect for the area. We did not believe it was a playground for tourists and nor would we encourage that; an ill-prepared traveller could die in the Sands. We had been there in the four coolest months. In mid-summer the temperatures rise dramatically, the winds blow harder and all the tracks become covered over. It would have been madness not to have had a Bedu guide to show us the way and to introduce us to his fellow tribesmen, and we had had the benefit of a huge support operation and the services of Taylorbase. Our advice to any visitors would be one of caution. We also wanted to be protective of our Bedu friends.

Taylorbase had served its purpose. It had been the 'field university' we wanted, where research, education and training went hand-in-hand and where proper dialogue between different disciplines actually took place. As we began to pack up, we all appreciated the contribution having such a mammoth camp had made. Not only had it been an efficient operations base, it had encouraged the close involvement of the people of the Sands and the Badiyah area. Taylorbase had acted as a *suuq* – a meeting place, a focus for interested parties, for government representatives, for travellers and for the local community wishing to share knowledge and news about the Sands. Our maps and satellite photographs on the wall bore many fingermarks and were now covered in felt-tip lines to represent our travels. Information was exchanged, both informally over dates and *qahwa* and also at the fortnightly workshops. These seminars grew in popularity during the project and by the end they had helped towards a proper sharing of results within the team and so contributed to our combined understanding of the area. Furthermore, it is quite clear that friendships forged between Oman and Britain at Taylorbase will continue and many of the team will return to Oman to pursue their studies.

One of the achievements of the project was this opportunity for so many to experience the Sands at first hand and make up their own minds what lay in store for the future. Taylorbase was used as a springboard to introduce interested parties to the area and so created fertile ground for discussion beyond the world of science. Many who might have remained on the hard top road now had sand in their boots. In that respect it brought key decision-makers closer to those on the ground who guard and care for the area. Our visitors' book listed over 500 who passed through Taylorbase and that had kept our kitchen, our guides and our vehicles busy. But no matter, for we could share our own excitement for the Sands – an area that had hardly been mentioned in literature and which was described in the local geographical text books as an 'inhospitable place where no one lives'.

In my view, this sharing process seems to be the key to all environmental issues. Collecting the facts is not enough. As that great educationalist Robin Hodgkin summarizes in his key work *Playing and Exploring*: 'These [unknown] worlds will only come to life if someone acts on them, plays with and explores them and then *shares* the resulting surprises.'

Is this not the responsibility that all geographers agree to when they have the privilege to explore new frontiers of knowledge? This charge will have to be taken more seriously by all environmental spokesmen whoever they may be. All those who bothered to cross the threshold of Taylorbase, past our ever watchful guard Ali Abdullah, will pass on their own comments and views of the Sands and will inspire others to come and take a closer look.

There is no doubt in any of our minds that this part of Oman, with its infinite variety of sand, is a remarkable example of a small, isolated, sand sea desert and that its regular dowsing of moisture from the night mists make it unique. As the summer winds began to cover our tracks, we wondered how many pieces of the geographical jigsaw we had discovered and, more importantly, how many pieces we would be able to join together. Any understanding of the complex processes that unite the earth, life and human components would be of value, but we must be clear in our descriptions and confident in our findings. Our statements might be of influence and so any reports had to be based on solid facts. Certainly there is no other body of sand like it anywhere in the world. As a sand sea it is a handy size – we were able to drive round it in two days. You cannot do that with either the Sahara or the Rub' Al Khali.

Small it may be, barren it is not. The most important feature of this land is that over 3,000 nomadic Bedu make it their home. 'We choose to live here,' Said reminded us many times during the project. For us, it had been a privilege to stay with Bedu families in the Sands, to join in their family life and their festivities, witnessing a way of life that made full use of the land and its plant resources. Life under the palm-leafed shelters is comfortable and healthy. Bed in the soft cool sand cures all aches and pains; stars above surpass any roof.

Coming to terms with this huge sandpit of infinite variety of colours and shapes had taken up every bit of the four months we had. For some it was still quite awesome but most of us were at home and happy to travel to all its corners. Our desert scientists could never stop describing the Sands in superlatives. They reminded us that it had been home for people for over 7,000 years and produced new-found stone tools as required evidence. In turn we confidently explained to anyone who would listen (often kneeling with sand falling through our hands), that here was the best specimen of an erg anywhere in the world.

Even for those unfamiliar with sand deserts, the variety of patterns and sizes of the dunes was obvious. The diversity of colours, the variation in vegetation cover from completely barren by the coast to the thick prosopis woodlands around Field Base gave some indication that the scientists' superlatives were not an exaggeration. With all examples in the handbook on sand dunes now ticked off, it was not difficult to believe that it was a sand sea that would become internationally renowned and which scientists from all over the world would want to visit. Its accessibility, being just two hours from Muscat, made it the 'living laboratory' we had originally described when we were trying to raise funds for the project so many months ago.

The whole team had enjoyed learning more about the geomorphology of the Sands and its new status as the site of the largest aeolianite field in the world was something we were all proud to mention. With not just one, or even two, but three old and now hardened sand seas lying underneath the Wahiba, we had the platform we needed on which to put the pieces of the jigsaw we were collecting.

The wadis that kept the Sands at bay were a story on their own. No one doubted their role having once seen the Wadi Batha in full flood. No wonder the Sands had never reached Mintirib. Occasionally, when the winds blew a little harder, some had passed the sleepy sentinels of the dry wadis, but the overall frame in which the sand was allowed to remain was as if finished with a knife – including the south-east corner where it meets the Arabian Sea. Having a picture of the Sands as a finite ecosystem made it easier to understand where the sand itself had come from and where it might be moved by wind or water today. We now know that all sand in the Wahiba Sands had originally come from broken-up rocks from the mountains in the north or from the ocean bed. Over 400 samples of sand in plastic bags back in Durham confirm this and are clues to the development of a sand sea. Most of these samples confirmed high calcium and magnesium carbonate derived from the shells of marine animals.

Many of the team spent time at Field Base, intrigued by the sand dune experiments there and the myriad technologies we had monitoring them to give a better picture of how the dunes move during different seasons and under different wind conditions. The dynamics of sand movement are always fertile ground for discussion. I think we were all astounded to learn that the crest of a dune could move as much as a metre in just twenty-four hours. Wooden pegs placed one evening would be lost the next. When there were erratic winds, during the January to March period, the dune crest just moved back and forth. During the harsher periods between March and December, the whole dunes moved forward and changed into broader, flatter shapes. At present our model seems to challenge existing theories, but that is not difficult as so little extensive monitoring of dunes to this degree of detail has ever been done. The work at Field Base continues, as our team have established a long-term monitoring programme with the Sultan Qaboos University. It will still be a few years before we fully understand the dunes but when we do, then, and only then, will we be able to control them. I believe that these studies in the Wahiba Sands have implications for the world.

Several topics grew from discussions at Taylorbase that linked the earth and life scientists. Geomorphologists worked well with our 'green' team. The sand team had begun to understand better how a dune held its moisture. The sponge-like quality of sand was necessary as rain might take up to three years to reappear. However, the sand was always moist under a top protective layer and some parts of the dunes were wetter than others. This was of interest to the biologists, who ultimately wanted to find the optimum conditions for planting and growing shrubs that would stabilize a moving dune.

The mists were also of interest to both. For all members, one of the most memorable experiences was the tremendous wetting we received on sleeping out on clear nights. Sometimes puddles would form in folds of the sleeping bag. We knew the mists were significant, but when it was confirmed that they on occasions deposited the equivalent of 0.5 millimetre of rainfall in the form of dew, it was no wonder our plant team became excited. You did not have to be a fully-fledged scientist to deduce from the dripping lichens in the prosopis woodlands that the mists were of influence – but now the quantities had been measured our belief that the animals and plants of the sands could survive without proper rainfall for months and maybe years was endorsed.

The biologists' findings concerning the sheer diversity of life in the Sands confirmed that deserts are always richer in life than they at first appear. As a desert kingdom it exceeded our expectations. Our score of 150 different plant species was certainly an improvement on the twenty-odd originally listed, although it had to be admitted that few scientists had had our opportunity to travel freely around the Sands. To find individual species never before recorded in Oman was exciting for our taxonomists. The discovery of the dew-drinking beetles that had only previously been documented in the Namib was as thrilling as that of the white-tailed mongoose hitherto unknown to inhabit that part of Oman.

As we ticked off an increasing number of animals it became obvious that the Sands were an important refuge for larger mammals, which in turn relied on a good supply of plant life or invertebrates. As our collectors confirmed, both were in sufficient abundance to sustain healthy populations of camels and goats, which in turn competed for space with gazelle, foxes and sand cats. Gazelle such as the white oryx have been hunted in the past and numbers in the Sultanate are very low indeed. They are relatively safe in the Sands and, if the ban on hunting is adhered to, there is a good chance that all future visitors there will glimpse these shy but majestic creatures darting behind the dune crests.

One of the great features of the Wahiba Sands are the prosopis woodlands. With roots deep into the sand, these dew-drinking trees forming woodland communities on the west and east of the Sands are home for a wealth of plant and animal life. You have to sleep among the trees to appreciate how very substantial they are, or even rest under them in the heat of the day. Many animals do, and so do the Bedu, who have several camps dotted amongst them. The trees provide this badly needed shelter, wood for cooking, timber for little huts, food in the form of edible fruit and leaves and fodder for livestock.

There is no doubt that our work on the prosopis, described as the 'Omani tree of life', will put this remarkable tree at the forefront of desertification issues. Its ability to grow where others cannot and to withstand continued harshness is still relatively unknown. We studied the diversity of its forms to identify the best types and so collect their seeds for long-term laboratory work. Using information about the best locations to plant such seeds in a dune and then protecting the young shoots from hungry livestock will give new life to the future of this tree. Specific follow-up work on the prosopis woodlands is considered high priority as we believe prosopis to be an effective plant for curbing the spread of desertification. We now know its requirements and what it can achieve when it has optimum conditions. Putting this tenacious plant on the map outside Oman and in other more mobile sand deserts will be an exciting future step for our biologists.

One of the recommendations that has already been passed to the Omani Government concerns the conservation status of the Wahiba Sands. Based on our initial findings, our team believe the area to show sufficient diversity to warrant it a site of scientific interest category. Mindful that Oman already has a plan for protected areas, our team suggested that there should be continued scientific scrutiny and that the Sands deserve the status of a National Resource Reserve at the very least.

There are of course severe limitations to short-term fieldwork. Four months is only part of the story. Working in the Sands for a full twelve months is now high priority for future research, although this in itself will be a mammoth operation. To secure information about the hot summer months, when even the Bedu lie low, will need very special logistics.

Despite this restriction, we did complete a vast amount of research in those four months. I was in awe of the team who could survive on very little sleep in order to monitor experiments going throughout the night or to make sure the day's data had been punched into the databases. The computer room was open twenty-four hours and the machines were rarely turned off. The compilation of data from thirty-two scientific disciplines all focused on the Sands is the very stuff that commissions such as the Brundtland Report are calling for. The severe lack of detailed knowledge about ecosystems and processes will become most apparent when politicians expect immediate answers to disasters caused by drought, by erosion, by deforestation, by flooding and by desertification. Few Prime Ministers or Presidents are yet ecologically conscious, but they soon will be as environmental issues climb the political agenda. When this happens, I hope that the value of expeditionary science as a means of creating a cell of research activity will soon be properly recognized as a tool for development.

No one on our team had been disappointed. Without exception all concerned felt that the Sands had relinquished many secrets during our expedition. The combined time added up to over forty years of research effort by the team and we intended to publish it within two weeks.

We had a strategy for sharing our results with a wide audience right from the start. All those who joined the team knew they had to work hard to ensure reports and information were produced fast. We were able to do this because data had been punched into the computers throughout the four months. Writing reports is an issue that affects all who work in the field. Many United Nations and government agencies have failed in the past because reports have sometimes taken several years to materialize – or, worse, never appeared at all. There was an easy solution. Passports and return tickets to Britain were handed in – to be returned once copy had been received and checked! Once back home, other pressures such as university posts and family holidays take precedence. Also, there was more motivation to write while the sand was still between the toes. Never again would the team be so in touch with the issues and points of information.

The end arrived all too soon. Richard Turpin, the expedition photographer, lined us up for the final team picture. There were mixed emotions. Soon we would be home with our loved ones; some had been away for five and a half months. But the bonds of friendship between the team members were also strong and it would be sad to leave the Sands and its people and see the team split up. Temperatures were over 40°C that day, an indication that summer was on the way and the Sands would soon be unbearably hot.

So on 1 April 1986 the whole team departed from Taylorbase and settled into the magnificent new campus of the Sultan Qaboos University for two weeks to discuss our initial findings with over 150 government representatives who came to workshops organized by Roderic Dutton. It was a valuable sharing time but, most importantly, every single member of the team typed up their own findings on to the computers and a 600-page report with maps and diagrams summarizing our initial results – entitled 'The Rapid Assessment Document' – was printed and bound and handed over to the Omani Government before we boarded our Gulf Air Tristar back to Heathrow.

We like to boast about it. We believe it was a breakthrough in publishing scientific results and a fitting response to the vision and commitment of all those who had given so much unselfish support. The value of the field back-up provided by the Sultan of Oman's Armed Forces was incalculable. The help given through the many ministries involved, and the Oman Co-ordinating Committee chaired by Colonel Khamis Mohammed Al Amry was, again, unprecedented in the history of such projects. The financial sponsorship by the eight Corporate Patrons and very many individuals was generous and unselfish. The total cost of the project was well within the overall target of £210,000. Many would argue that international agencies around the world are spending several hundred times more money on their fieldwork with much less to report in the long term.

However, the final epitaph of the project will be secure in the Sultanate's own scientific publication – the *Journal of Oman Studies*. To be published as a special report by the Office of the Adviser for Conservation and the Environment, over fifty scientific papers summarizing the results of the expedition will be bound in one monograph. This will be our statement about Sands and Sand Sea Deserts, which will have an active life offering solutions to desertification issues and be a platform of knowledge on which to build. Then and only then, can the team relax their commitment to the Sands. But it is Roderic and his team at the Centre for Overseas Research and Development who now shoulder this burden and it is with his permission we include his Scientific Overview and Conclusions (see Epilogue) that will appear as an introduction to the final report.

The area as a whole is in good hands. In Sultan Qaboos, Oman has one of the 'greenest' statesmen of any country and his environmental crusade is already well established. The support provided by His Majesty and the guidance given through his Conservation Adviser, Ralph Daly, was evidence of this. As environmental issues top political agendas, Oman will show that it has been leading the field by integrating development and conservation for many years. Sultan Qaboos has been wise not to make mistakes which have beset other nations keen to rush ahead too fast. Many of the team believe that Oman has a number of lessons to teach the world and there is no doubt that the Royal Geographical Society in its endeavour to promote geographical knowledge for conservation and for development found an important ally in Sultan Qaboos and his government.

Four months in the Sands leaves an indelible mark; it had changed the lives of the team. Few could now speak of the area and its people without enthusiasm and we were all protective of our new-found friendships. Those on the team who had never been to Arabia Felix had shared a fear of causing offence by upsetting established traditions. For some, indeed, this visit had been the very first introduction to Arab people. It had been a rich introduction and no one took for granted the privilege of having had such an experience.

The links between Britain and Oman are strong but it was only by using the opportunity to work closely within a community that we learnt just how strong. There were constant reminders of close ties throughout Oman. When Dhiyab and his team went from house to house in Mintirib they met the head of one family who was responsible for keeping the town dustbins clean, a job he did with pride. His sons tended date gardens. On enquiry about the daughter, Dhiyab was interested to learn that she was studying political sciences at London University.

My most treasured memory of the project is that of my trip with Said and Dhiyab, from which I still bear the camel scars. It put into perspective my own understanding of the true desert life and its influence in Arabia as the source of hospitality and friendship. There is a long British tradition of love for Oman and its peoples and we are now proud to have been a part of it; the relationships our team shared with characters like Said Jabber, Abdulhakim and Khalifa will be talked about for ever.

During our last days in the Sultanate we enjoyed a wealth of Omani hospitality and many people generously entertained the whole team. We, in turn, hosted a number of receptions to thank the several hundred people in Muscat who had helped us achieve so much in so short a time. The last of these was a sit-down dinner arranged at the Hotel Intercontinental Muscat to which the Deputy Prime Minister, His Highness Sayyid Faher bin Taimur, was our guest of honour. The management had laid on a full banquet with the compliments of the house and no expense had been spared to offer the best cuisine in Muscat. The hotel is a large, fort-like building which, once inside, has a covered courtyard with glass lifts rising and falling in the centre. Gardens and fountains create an oasis from the harsh heat outside. It was a most unreal setting; magnificent but like something seen in dreamland.

The next day the team flew back to Britain. Roderic Dutton and I remained to say farewell to many others in Muscat, but on the final day we couldn't resist returning to Badiyah to say farewell to Said Jabber Hilays and his family in their town house at Mintirib.

Said invited us into his *majlis* and we squatted down on the floor, leaning against cushions. It had recently been painted and had an electric fan installed. Said was pleased to see us and asked his wife to prepare an enormous lunch while we brought him up to date with our news of how things had gone in the capital. We argued that we were not hungry but he insisted and as we had driven down especially to see him we could not easily have said that we had other pressing matters to attend to. For the first time in four months I could relax.

Salam Said his eldest and Salim Said his number three son served dates and coffee. On the wall were one or two photographs of project members and a map of the Sands. We did not have much to say and sat in silence, not believing that we would soon be on our way home. Said, his *masarra* cocked at a cheeky angle, told us not to be sad. The Bedu would be here when we returned and they would continue to be the Sands' guardians. Said had a genuine interest in establishing a service to scientists and travellers alike. He was keen to set up a business to help anyone who wanted to work in the Sands.

After a large chicken and rice lunch served on an enormous tray, Said regarded us with no perceptible regret. Our friend resorted to his familiar Bedu look, aristocratic, friendly and confident.

'The Bedu will always be pleased to see you,' he assured us.

Roderic and I thanked Said and bid farewell to his family and seven children. Said then plucked at my sleeve and took me aside.

'Don't forget you are now honorary Bedu and you must tell people about the Wahiba Sands. Come back with your family.'

'Jazaak allaah khair – may God reward you well.'

'As salaam 'alaikum.'

''Alaikum as salaam.'

'Ma' as salaamah. Ma' as salaamah.'

EPILOGUE

Overview and Conclusions

Roderic W. Dutton

CENTRE FOR OVERSEAS RESEARCH AND DEVELOPMENT
University of Durham

Introduction

The principal characteristics of the Sands, earth science, biological and human, are examined in depth in successive chapters of this volume[1]. The introductory sections to this opening chapter, therefore, are limited to some brief definitions of name and study area, a list of principal Wahiba Sands characteristics, a statement of project objectives and an outline of the fieldwork programme. The overview explores some of the main themes of the research programme that run through different groups of the papers, and shows how each piece of work is related to and dependent on the others to produce conclusions arising from a full multi-disciplinary synthesis. Research without effective communication, however, has no value. In addition to publishing this monograph, therefore, the Project has made every effort to communicate with as wide an audience as possible, and this work is also outlined here. General conclusions are categorized under the headings research, education, conservation, development and planning.

The Study Area

NAME

Amongst the many areas of scientific discussion and controversy generated by the Project was the very name of the sand sea itself. At the outset it appeared to be clear: internationally, on all maps, and also within Oman itself beyond the Sharqiyah region the sand sea in question is known as the 'Wahiba Sands', hence the title of the Project, the 'Oman Wahiba Sands Project 1985/87', and the name on all project reports.

However, we now know, as a result of close contact with many people living in the Sharqiyah, including contact with many members of the Al Wahibah tribe, that the term Wahiba Sands is not one commonly used locally. Indeed, there does not seem to be any commonly used name for the sand sea as a whole, though there are many names for different parts of it. Moreover, the use, by Project members, of the term Wahiba Sands caused confusion and some annoyance partly because a tribal name was being applied to a geographical entity and partly because the name, used in this way, appeared to be implying exclusive rights to the Sands by the Wahibah tribe. Project members therefore, in practice, used the entirely neutral name, the 'Sands' for the sand sea and this is the convention that in general we follow in this volume. But as His Majesty Sultan Qaboos bin Said has sanctioned the continued use of the term Wahiba (as applied to the sand sea) internationally the name Wahiba Sands appears in the title.

LOCATION

The Sands lie in the eastern region (the Sharqiyah) of Oman between approximately 20′ 45″ and 22′ 30″ north, and 58′ 30″ and 59′ 10″ east (Map 5.1, p. 64). Their maximum dimensions, as estimated by Jones, Cooke and Warren, are approximately 195 km north-south and 85 km east-west, and their surface area 9,400 km^2. To the north and west of the Sands proper lie areas of only thin and broken sand cover referred to in this volume as the 'Peripheral Sands'. Their extent is estimated at 3,500 km^2.

The Wadi Batha runs along the north and eastern margin of the Sands separating them from the Eastern Hajar mountains and the Ja'alan. To the west of the Sands are gravel plains in which lie the Wadis Matam and Andam, draining approximately north-south. The south-eastern boundary of the Sands abuts the Arabian sea.

THE SANDS MARGINS

The Project includes within its scope of work both the Sands and the Sands margins. To understand the Sands, from all standpoints (earth science, biological and human) a study of their margins had to be considered as an integral part of the programme. The following three examples illustrate the point. Firstly, the Sands which compose the sand sea are wind and water borne sediments originating from both terrestrial and marine margins. Secondly, it has been suggested that the Wadi Batha is a 'gene-pool' for plant seed later dispersed into the Sands. Thirdly, many villages lie at the Sands margins and interdependences between villagers, fishermen and bedu are as varied as they are important. Recently new influences on the Sands and on their resources and inhabitants have reached the margins, and the Sands themselves, as direct and indirect consequences of oil wealth and government intervention.

PRINCIPAL FEATURES

The northern half of the Sands (the 'High Sands') is characterized by north-south orientated megadunes, up to 100 m in height, with crests about 2 km apart and separated by broad swales. The megadunes are relatively stable, and their general topography and orientation are not affected by the annual cycle of wind regimes. In the southern half of the Sands (the 'Low Sands') the relief is lower and the orientation of the dunes much more varied. These dunes are typically active with much mobile sand responding to wind changes on seasonal and daily cycles. Many of these dunes are soft, making travelling by vehicle slow and sometimes hazardous. Jones, Cooke and Warren give a detailed terrain classification of the whole area and describe the morphology of the 33 terrain units recognized.

Special vegetation features include the *Prosopis* woodlands along the eastern and western margins of the central Sands, and beyond the Sands in the south-west. The *Prosopis* woodlands are an important forage resource and create shaded, pleasant conditions which attract large numbers of bedu to live amongst them. In the remaining areas of the Sands vegetation cover varies from quite dense (especially after rains) to non-existent. No significant area of agriculture is found within the Sands, and permanent settlements are restricted to the Sands margins. Bedu fishermen occupy small settlements of simple construction along the coast.

Project Objectives

SCIENTIFIC OBJECTIVES

The central scientific aim of the Sands Project is:

– A study of the early development of the Sands, their ecosystem and the impact of recent change.

From this aim were formulated, consequent on the mapping phase of the Project in January/February 1985, the five scientific objectives which formed the framework for the research programme and this scientific monograph.

The five scientific objectives were:

- *Sedimentary and geomorphological history*: The reconstruction of the late quaternary environment of the Sands concentrating on three elements; raised channels, aeolianite, and evidence of early human occupation.
- *Sand movement, moisture and vegetation*: Measurement and development of models of dune movement, and soil-moisture budgets. These were to be related to the distribution of plants within the Sands providing information for vegetation management and sand control programmes.
- *Biological resources and range management*: Description and collection of the flora and fauna, including special studies of the *Prosopis* woodlands, the range, invertebrates, amphibians, reptiles, birds and mammals. This, together with a study of the use of the biological resources by people and their livestock were to provide information relevant to plans for improving resource management.
- *Indigenous communities and their inter-relationships*: Study of the productive activities of the area including the social organization of production, and examination of the networks of exchange within the Sands and the means of regulating access to resources.
- *Oil wealth and development*: Examination of changes in family income and consumption related to oil-derived wealth and resulting demographic changes, and effects of settlement patterns and land use, and also a study of the evolving role of government services, and of the role of government organizations in promoting economic development.

Where relevant, emphasis is placed on the inter-relationships which link the five areas of study given above, pointing out the continual and often rapid changes taking place in many aspects of the Sands systems under examination as a result of the impact of climatic, geomorphological and socio-economic processes. (Figure 1).

APPLIED OBJECTIVES

Oman is primarily interested, at this stage of its development, in obtaining information about its natural resources and using it to address the requirements of resource management problems and of opportunities for development. Oman's requirement is for effective, sustainable development, making the best longterm use of its resources to the benefit of its people, without threatening the ecological system on which that development depends. Helping Oman realize this ambition was therefore a main thrust of the Sands Project: producing relevant facts and presenting them in a

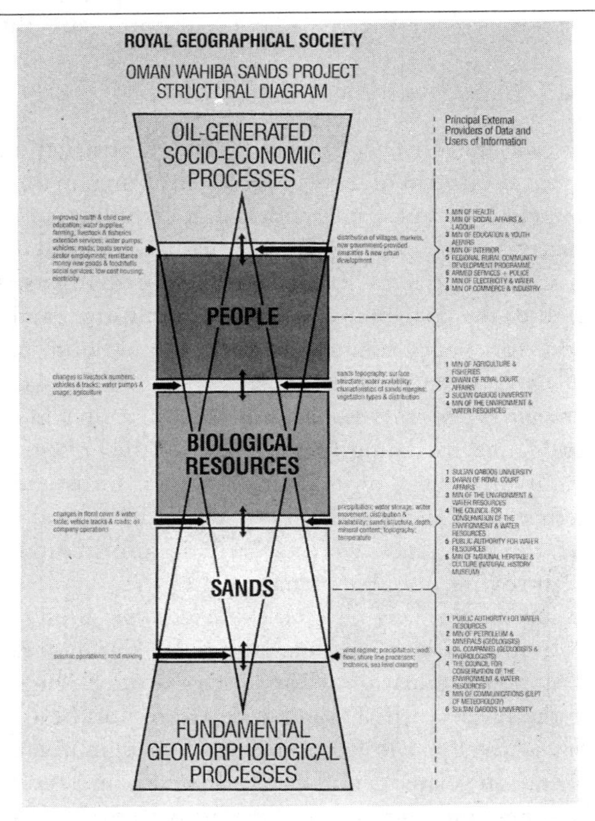

FIGURE 1. Project structural diagram (Omani organizations as they existed in 1985/86, when the Project was in the field).

way which will best facilitate their usage by the authorities in Oman to ensure development that is environmentally sound.

However, neither this monograph nor previous reports attempt to provide a set of 'project proposals' or 'feasibility studies' to match either development or conservation needs. It has not been part of the Project brief to provide the details of scope of work, nor of manpower requirements, nor of costs that would be required for such proposals or studies. But some conservation problems and opportunities for development are given a closer definition in some of the papers, and suggestions made for tackling them. Success in following up these suggestions will depend on continued and close co-operation at Ministry level, and on additional fieldwork to suit the requirements of specific projects.

EVALUATING TECHNOLOGICAL AIDS TO FIELDWORK

The Project tested the extent to which survey technologies and computers, and Landsat and other imagery could help with fieldwork and report writing; the extent to which in practice they could increase the amount of information available to the Project, hasten and make more exact the collecting of data, enhance data analysis, and facilitate rapid report writing including the computer production of text illustrations.

Thus four members of the Project had special responsibility for surveying, computers, Landsat image analysis and computerized geographic information systems.

A JOINT PROJECT

The Project, from its inception, determined to work with (as well as for) its Omani hosts. At Government level this ambition was achieved by the Oman Co-ordinating Committee on which were represented all the Ministries, the Armed Forces and the Sultan Qaboos University.

It was also recognized by the Project that a great deal of information about the region was held in Oman. Firstly, of course, since 1970 the government of Oman has been actively undertaking surveys and feasibility studies in the region (Figure 1). All ministries have been involved, as have other organizations such as oil companies, the Public Authority for Water Resources (PAWR) and the Public Authority for Marketing Agricultural Produce (PAMAP). In practice, given the Project's particular range of interests, it benefited most from reports and other information kindly made available to it by the Ministry of Agriculture and Fisheries, the Ministry of National Heritage and Culture, the Ministry of Social Affairs and Labour, PAWR (now the Council for the Conservation of the Environment and Water Resources) and PAMAP. Their contributions to this Sands research programme, and all others, are gratefully acknowledged. Secondly, the people living in the Sands and on their margins have, of course, a profound knowledge of the area. Many informants most openly and helpfully provided a wealth of information about local resources and the indigenous economies (pastoralism, fishing, farming and crafts) and how they are being affected by recent change. Others acted as interpreters and as guides throughout the area.

The Project was also greatly strengthened by close association with researchers who work for or were sponsored by Omani organizations. They made major contributions in the fields of geomorphology, archaeology, dew measurement, animal collecting and identification, plant taxonomy and economic change. All have provided papers for this monograph.

Many other people from Oman, or working in Oman, made invaluable contributions to discussions in the field workshops, in the discussion workshop held at Sultan Qaboos University (SQU) at the beginning of April 1986, and at the symposium held at SQU in April 1987.

But apart from the scientifically essential requirement of the Project to work closely with people long familiar with Oman, a joint Project was also important because responsibility for the management of the region resides with Oman. The people of the region may be regarded as the local guardians of the Sands, and their exploitation of the local resources has traditionally been tempered by the knowledge that to misuse them courts disaster. But in recent years the private wealth of the Sands inhabitants has increased sharply, and as most of this is remittance wealth it will have lessened the bond of dependence between the people and their local resources and therefore, as the Project discovered, some of their concern for the longterm well-being of these resources.

Additionally today, and broadly speaking since the start of oil exports and the accession of Sultan Qaboos in 1970, the government of Oman has become much more directly involved in the region through the introduction of many amenities, utilities and services and is also, therefore, centrally involved in management.

Fieldwork Programme

The overall fieldwork programme comprised: the mapping phase, the main phase and the final phase.

In the mapping phase (4th January to 7th February 1985) a team of seven scientists undertook a preliminary mapping and reconnaissance survey of the Sands. With the help of a very high level of support from the Sultan of Oman's Armed Forces, notably land transport from the Coast Security Force and helicopter and fixed wing flights from the Sultan of Oman's Air Force, a complete overview of the Sands and the sand sea margins was accomplished. Many sand samples were taken for subsequent analysis in Britain, and collections of plants and invertebrates were started. Some discussions and interviews were held with inhabitants of the region; pastoralists, fishermen, farmers and villagers.

The mapping phase also provided the opportunity for continual discussion, between team members and with locally based scientists, about the work in hand and about the main phase programme. These discussions gave rise to the first formulation of the five scientific objectives (given above) to be attained by the overall field programme.

The main phase (2nd December 1985 to 15th April 1986) comprised a team of 32 scientists, including associate members. In the field the team operated out of a substantial 40 bed main camp ('Taylorbase') at Mintirib, a Field Base at Qarhat Mu'ammar and a mobile base which moved from site to site mainly in the south-western area of the Sands. Once again the programme benefited enormously from extensive and very varied support provided by the Armed Forces, in close liaison with HQ SOLF, including substantial logistic support notably with land and air travel within the Sands. Additionally, divers from the Sultan of Oman's Navy collected marine sediments and surveyors from the Sultan of Oman's Artillery assisted the Project's own survey team.

The research team divided into three groups to study, respectively, earth sciences and water, biological resources and the people. The work focussed on the five scientific objectives, and a preliminary write-up of the results was given in a Rapid Assessment Document (Dutton, 1986).

The final phase (or phases) of the fieldwork were undertaken by individuals returning to the field at different times up to the end of 1987. They made comparative studies at different seasons and under different climatic conditions to complement the work previously undertaken.

Overview

EARTH SCIENCES

Terrain and geology: The terrain classification by Jones, Cooke and Warren has already been sited. This classification of the whole area was in part based on the examination of Landsat imagery. At a higher resolution McBean has also analyzed Landsat Thematic Mapper (TM), and SIR-A and ground radiometer data, to provide detailed land cover information of some selected areas of the Sands and their margins. Work on the Eastern *Prosopis* woodlands, for example, shows that the densest areas tend to be at the western edge. In the near future, and with the aid of more ground truthing, it is hoped that the number of trees and total biomass might also be estimated. Such applications of image analysis should make it a powerful aid to environmental management; monitoring through time the cumulative effects of, for example, problems of land reclamation schemes, population pressure, dune movement and salinization.

Kay (GIS) stresses that a major application of the Geographic Information System (GIS) programme is in the research and use of an area as a training ground for the rest of a Landsat TM image scene. In the case of the Wahiba Sands Project several datasets were included in the GIS database including those covering earth, biological and human sciences, together with Landsat and survey data. Having classified the GIS area (Badiyah) it was seen that when using the classification to analyze a subscene of similar topography the match was imperfect; the subscene analysis indicated more built-up area and more wadi channel area than was actually the case. However, this mismatch is simply a problem of refining the technique; the point is demonstrated that the GIS/Landsat image combination is a valuable tool for land evaluation, monitoring, assessment and planning.

A description of the geological setting beneath the Sands is given by Glennie, often inferred from the rock units that surround the Sands for lack of data from deep boreholes within them.

Palaeoenvironments: Beneath the surface dunes of the Sands a sedimentary rock of special palaeoenvironmental interest is the aeolianite (cemented aeolian sand) studied by Gardner. It appears to underlie most of the Sands and extends at least 3 km offshore, and is the most extensive continuous deposit of aeolianite known in the world. In many areas the aeolianite outcrops, and the structure, orientation, depositional layering and wind deflated terracing, together with associated stranded shorelines provide many clues about the quaternary history of the Sands. Gardner identifies four phases: periodic accumulation; deflation and working of aeolian deposits inland (to the High Sands); a relatively humid interval in the Holocene; and finally an arid period of more rapid recent erosion. Gardner also suggests that the three main controls are sea level, climate (wind, precipitation and humidity) and height of water table.

In the area of raised channels to the west of the Sands Maizels identifies five phases of development: initial development of the piedmont alluvial plain; prolonged fan aggradation with rising water table; a renewed period of fan sedimentation; sudden and dramatic degradation of the fans and extensive deflation of the interfluves and the gradual exposure of the palaeochannels; and the modern wadi system. A study of the flash floods in the Wadis Andam and Batha of February/March 1986, by Maizels and Anderson, is indicative of the fluvial forces involved in recent times and of their geomorphological consequences.

The earliest evidence found by Edens for human occupation of the Sands dates to the humid Holocene identified by Gardner. One of the stone-tool industries, characterized by points and commonly associated with geomorphic evidence of standing water, suggests a hunter/gatherer population in the Sands. From the same period another stone-tool assemblage was found in the Andam-Halfayn drainage basin and is similar to assemblages found throughout the mountain areas. From later periods significant evidence of occupation dates only from the late 16th century to the present time; glazed and plain pottery is found on the coast and indicates modern human activity along the coast to be of only recent origin.

Sand: The uncertainties about sediment types, sources and transport mechanisms expressed by Allison are indicative of the complex and ever changing balance of geomorphological processes that have shaped, and continue to shape, the Sands. Roughly half the sand in many locations is quartz, 30–45 per cent in different locations is carbonate of marine origin, and 10–21 per cent is of ophiolitic origin. These materials have been brought into the Sands from the Oman mountains and from the nearshore continental shelf by aeolian, marine and fluvial processes, and have also been reworked and re-transported within the Sands as, for example, in Gardner's second phase of the Sands development.

Such sand movement, within present-day dune fields, is under examination by Warren (Dynamics) with the assistance of methods of land survey and photogrammetry refined and reported on by Kay (Techniques). Warren reports that although the gross morphology of the dunes he studied in detail at 'Urayfat was not affected by the light winds of January to March (1986) it was profoundly affected by the strong and constant winds from March to July. Indeed, he estimates that these dunes are being propelled inland at an average 10m annually, and left the coast some 4 ka ago. This rate of advance, and the strength of the wind forces which cause it, have very strong practical implications for the builders of roads and other installations in the path of the mobile dunes.

Along the northern margin of the Sands, however, earlier fears about 'desertification' may have been exaggerated. Brunsden, Cooke and Jones examine the conclusions of a previous report which indicated that parts of Badiyah were in danger of being over-run by sand blowing from the main dune fields. They give a preliminary definition to the paths of sand movement and to the forces controlling the movement. However, they conclude that although sand is seen in some fields at the distal end of the *falaj* systems in some villages, it builds up only after the fields have been abandoned for other reasons. Also, they note that local techniques of controlling sand, by erection of barriers or physical removal, are effective and adequate to the need.

Nevertheless the problems caused by sand movement give a practical justification for the improved theoretical modelling of dune movement which Warren (Dynamics) calls for. The need for improved knowledge and understanding is also a theme of Warren's (Dunes), and he states that: '. . . hypotheses on how dunes grow can be better tested here than anywhere else on Earth'. He describes the great variety and complexity of the dunes, examines them with reference to current theories of dune formation and speculates about the recent geological history of the area. But he says of the hypotheses he mentions that much more data will be needed before any can be accepted, and he lists eight fields of study that would be both feasible and valuable to undertake.

Water: The work of Gardner, of Maizels, of Warren and of Allison underline the key role that water has played (and continues to play) in the evolution of this arid sand sea. And water is, of course, essential to the plant and animal life of the Sands and to the presence of man. An understanding of surface hydrology is, therefore, of wide significance. Agnew's interesting findings, which benefited from the coincidence of rainfall, include: after rain, a rapid drying of the surface 5–10 cm of sand within a month which insulates the subsurface against evaporation losses and prolongs water availability for plants; an accumulation of water at depth in interdune depressions as a result of lateral seepage for some time after rainfall which could make the depressions attractive sites for colonization by plants; and an anti-wetting property of soils in the *Prosopis* mounds which will favour shallow rooting plants.

Less dramatically obvious than desert rain but of more frequent occurrence, dewfall has been shown to play a role in sand movement, in plant growth and in supplying the water requirements of some animals. The pattern of dewfall was examined by Anderson at sites in and around part of the *Prosopis* woodland. There are clear relationships between dewfall and location within the woodland and within individual tree canopies. The results are preliminary, and it is therefore difficult to interpret the mechanisms involved. However, as the dewfall can be surprisingly high and as the potential value of dew as a water resource for *Prosopis* is shown by Brown and strongly suggested for other plants by the bedu, additional research will be of undoubted practical value, particularly when siting new forest plantations. Away from the *Prosopis* trees Agnew and Anderson found a strong relationship between the duration of saturated atmospheric conditions and dewfall. At the top of dune slip faces Büttiker and Anderson have shown that dew-drinking beetles obtain sufficient moisture from dew to supply their needs.

An initial picture of the subsurface hydrology of the Sands is presented by Jones, Weier and Considine from their analysis of the data from 33 holes drilled in the Sands in 1984–5 to test groundwater potential. They revealed an alluvial and an aeolianite aquifer located, respectively, in the north-east and east of the Sands and supplied, for the most part, by water flowing underground from adjoining areas. At best these aquifers yield 10–25 l/s in pump tests. In the north-east border of the Sands the water is of good quality, and in the east in general it is of medium quality. There is, therefore, some potential for developing the water resource for municipal or agricultural usage but this poses profound ecological problems, and usage should be constrained at least until more is known about the aquifer size and about its recharge, storage and discharge system. Overuse of the water could, amongst other things, lead to the death of the trees in the Eastern *Prosopis* woodland, as Brown forewarns.

BIOLOGICAL RESOURCES

Plants and their uses: About 180 species of plant have been identified by Cope who comments that although the flora is not rich it is remarkably diverse, including one species new to science. Some of this diversity stems from the fact emphasized by Cope that phytogeographically the Arabian Peninsula sits on a crossroads drawing its component flora from many regions. But only about 30 species were found in the Sands proper. The remainder of the plants identified by Cope were from the marginal areas. Cope states that it is the marginal areas 'from which the gene pool in the Sands is derived', thereby indicating the biological importance of the margins to the central Sands. Munton (Vegetation and Forage) is able to describe only 15 plant species as having some significant forage value, including the species found predominantly at the margins.

Munton (Overview) found some significant associations between the major plant species in the Sands. He also noted that whereas *Heliotropium kotschyi* and *Euphorbia riebeckii* are found in firmer sand, *Calligonum comosum* and *Panicum turgidum* are able to colonize the softer and more mobile sands. Warren (Notes) uses this information, and his own observations and interpretation of air photographs to suggest a vegetation succession as active dune fields migrate northwards. The mobile sands are colonized by *Calligonum* but as the sand is blown from around the *Calligonum* they are left stranded and in decay on large nabkhah (mounds), leaving the firmer, more subdued sand to be colonized by *Euphorbia* and *Heliotropium*.

Prosopis woodlands: The ecological importance of the *Prosopis* woodlands is emphasized by Brown who lists a large total of plants and animals (see Cope, Büttiker, Gallagher etc.) which live in association with these trees which provide cool, moist and wind-sheltered shade, soils richer in nutrients than elsewhere, and leaves, flowers and fruit as a food source for wild fauna and domestic stock. As Webster (Pastoral Ecology) says: 'It is in the wadis [*Prosopis* areas] that the interaction of man, domestic animals, wildlife and natural vegetation is most intensive. It is in the interests of all these elements that a balance be maintained.'

But in some areas the balance is being lost. According to both Munton and Brown, the northern third of the Eastern *Prosopis* area is threatened by heavy browsing of all stock and by pruning, and needs protecting.

But in spite of the heavy browsing and lopping of *Prosopis* trees, and in spite of the near total lack of regeneration from seeds because of predation and browsing, shortage of water is probably the major longterm threat to the *Prosopis* woodlands. Most of their water comes from the water table and as the trees lack common xerophytic features and appear, from Laurie's work, to transpire freely during the day, a drop in the water table below the roots would kill the trees. Such a drop in the water table is to be feared because new wells, equipped with diesel pumps, continue to be sunk, and because serious consideration is being given to ways in which the recently discovered aquifers, mentioned by Jones, Weier and Considine, might be used.

Invertebrates: Büttiker W. and S. (Invertebrates) report that some 16,500 invertebrates from 31 taxonomic groups were collected during the Project. These are being identified and, where new, described and the results will provide material for about 40 papers including those in this volume. When the work is completed, improved and new distribution maps can be drawn and we will have a better understanding of Oman's relationship with its three major neighbouring zoogeographical regions. Ecological studies, and the greater understanding that the Project has given us of animal-plant-habitat associations, will allow local plans for conservation and development to be made on a more rational basis.

More detailed habitat and ecosystem studies will also allow vectors of human and animal diseases to be controlled more effectively and with minimal danger to other species of animal and to man. A number of potential disease vectors and other invertebrates that can cause problems to man have been identified. Lane and Büttiker, for example, report that sandflies, including those known to be vectors of leishmaniasis in Saudi Arabia, are present though fortunately uncommon. Blackflies, which are normally associated with flowing streams and are vectors of river blindness have for the first time been found in Oman, and identified by Crosskey and Büttiker. The disease is known in North Yemen and Asir (Saudi Arabia) and a careful examination of potential breeding sites in running water in Oman must be undertaken before its occurrence here can be discounted. Popov describes the locust and grasshopper fauna of the Sands as not rich. However, mindful of the fact that Oman was the initial source area of at least two locust upsurges in the past four decades, Popov emphasizes the importance of identifying and surveying breeding areas, and controlling any gregarizing populations. A species of eye-frequenting moth was observed in Oman, for the first time, by Büttiker W. and S. (Ophthalmotropic Lepidoptera) on feral donkeys in *Prosopis* woodlands. The Büttikers recommend that these eye-parasites should be investigated as possible vectors of livestock diseases such as rinderpest, FMD, PPR and others. Chotani records 14 species of termites found in Oman (including two new to science), though not all from the Wahiba Sands.

Other invertebrate groups so far reported on cause little or no harm to man but indicate more of the diversity of life forms and the complexity of the ecosystem. A short account of the distribution and habitat of 28 ant species from the Sands and their margins is given by Collingwood including some normally associated with man-made environments rather than with deserts. Schneider lists 17 species of dragonfly, including four new to Oman, and discusses their Afrotropical associations.

Wiltshire, writing about 57 moth species of which 55 are larger moths, also discusses their zoogeographical links, and he makes important points about their conservation and control that are true for other groups as well. Of the 57 species only two are serious pests; one (*Arenipses sabella*) on dates and the other (*Spodoptera exigua*) on cultivated crops. Most of the other moths have a completely innocuous but important role. They feed:

> ... the whole life-community, and constitute therefore a resource that should not be overlooked. Any control measures that may be planned against the pests should not be indiscriminate, as too often such campaigns have caused general havoc in the community.

Wiltshire also emphasizes that:

> The only way to control the Sands Lepidoptera is to conserve the vegetation and to minimize use of general insecticides.... The use of poison-sprays should be avoided unless carefully supervised by a responsible entomologist aware of the useful role of lepidoptera in the life-community.... [If care is not taken] ... all that will remain of the [moth] fauna ... may be a few migratory pests which no amount of spraying with insecticides will completely exterminate.

Fish, molluscs, reptiles and amphibians: Fish are, perhaps, an unexpected discovery even from a wadi on the margins of the Sands, at least to those unfamiliar with the Omani *falaj* (Gabriel, Agriculture). However, three species of freshwater fish, including a new subspecies, found in the Wadi Batha are discussed by Krupp. He stresses the point that aquatic habitats in arid zones are usually unstable and easily disturbed resulting in the eradication of this part of the fauna. Eight species of land mollusc are listed by Mordan including two reported for the first time in Arabia. Ferrara and Taiti list five species of terrestrial isopod, including one new one. Gallagher and Arnold record 28 species of reptile and amphibian of which two are new records for Oman and an additional 16 new to the Sands area. This fauna includes a lot of sand-adapted specialists.

Mammals: The carnivores of the Sands, as studied by Linn, include the wolf (*Canus lupus*), two species of fox (*Vulpes rüppelli* and *V. vulpes*), wild cats (*Felis sivestris*) and two white-tailed mongooses (*Ichneumia albicauda*) apart from feral dogs and cats. Linn explored the ways in which the carnivores survive in this hostile environment. Apart from the feral animals, whose control Linn believes is desirable, the carnivores include generalist species (such as the Red Fox) not able to penetrate the heart of the Sands, specialists such as Rüppell's fox which flourish in sand deserts, and the wolf which is probably a vagrant.

The Rüppell's foxes are, therefore, the most interesting of the carnivores in terms of desert specialization. Their normal food is composed of the larger insects and the small mammals already mentioned. Linn was able to radio track two Rüppell's foxes and thus make some interesting observations on their behavioural ecology including the fact that 400 ha of sand desert is required to support one 1.5 kg fox. This result indicates that the Sands are relatively well stocked with fox food, at least in comparison with the neighbouring gravel plain of the Jiddat al Harasis where a similar fox has been shown to require a sixfold increase in area. Of special interest and importance is the marine margin. Work near the village of Quhayd revealed a density of foxes surprising for an area of desert notably bare of vegetation and therefore of the normal fox prey already mentioned, surprising that is until it was realized how rich the shore is in naturally occurring prey and carrion and in fish and fish waste near the fishing villages.

Small mammals, some of which are fox food, were also found in the study area and examined by Gallagher and Harrison: pipistrelle bats, jirds and gerbils in the Sands, and these and others including rats and mice at the margins. Though other species were indicated by their tracks the authors comment on the apparent depauperate nature of the species diversity. Additional studies are recommended to search for other species, to investigate behavioural strategies, and to make comparisons with other deserts.

Birds: Gallagher has prepared a systematic list of 115 bird species of which six are the most southerly and one the most northerly records in Arabia, and one is the first breeding record in Oman. A number of factors make the Sands less hostile to bird life than might, superficially, be expected. First of these factors is, once again, the marine margin (53 bird species), whose summer upwelling of nutrient-rich cold water and whose onshore humidity stimulate the production of food for the birds. A special feature of the coastline is the mudflats and sabkhah of the Barr al Hikman. A second factor is that the Sands are relatively small and the north-south path across them is indicated by the *Prosopis* woodlands which also form a habitat in which 29 bird species were found. Only 21 species were found in the central Sands.

Diversity: The low plant diversity in the Sands proper is matched by the low total of small mammals found (three species) and by the single species of ruminant. A number of small mammalian predators was found, but only the Rüppell's sand fox is a resident in the heart of the Sands. The majority of the birds are associated with the *Prosopis* woodlands or the coastal margin, and whilst reptiles, amphibians and large numbers of invertebrates were collected many of these were also of the marginal areas. Munton (Overview) sites several factors which may explain this paucity, including the presence of man. Man continues to hunt the Arabian gazelle, and his 26,000 head of goats which browse following rain will have altered the species composition of the vegetation. The margins support a much greater diversity of life. Many species, including the majority of the bedu and their livestock, find refuge in the margins and make only opportunistic use of the central Sands.

ECONOMY AND SOCIETY

Vehicles: The motor vehicle is perhaps the most potent symbol of social and economic change in the Wahiba Sands; in particular the red, balloon-tyred, Toyota pick-up trucks favoured by the bedu. The great growth in the number of vehicles used in the Sands and their margins is one of the more visible consequences of oil wealth and a powerful stimulant to additional change. Vehicles, and the new network of roads on which they run, have, for example, put the fishing community into direct economic contact with the Gulf states. Vehicles carry an increasing range and quantity of goods from the Capital area into the villages adjacent to the Sands, and from the villages into the Sands.

Webster (Gabriel *et al*) states that vehicles 'have transformed the lives of the bedu in many ways'. They have replaced the camel for riding and for the transport of goods, and placed even the north and south extremities of the Sands within a few hours journey time of each other. Because they allow the rapid and easy transport of goods they encourage the accumulation of possessions. Water, and purchased feed, can be taken to the animals and weighty building materials can be carried from the villages to what, in some cases, begin to look like permanent homes; homes that are both cause and consequence of decreasing mobility.

Future of pastoralism: But in spite of the much closer links now forged between the Sands and their margins by the vehicle and all that it symbolizes, Webster (Economy and Society) still argues that 'pastoralism is virtually the only economic resource of areas such as the Wahiba Sands'. Munton (Evaluation and Management), however, questions this view, pointing out that the area has a tourist potential. Munton's ideas underline the importance of the concept that from the viewpoint of national development strategy the bedu should be regarded not primarily as herders of camels and goats but as people who uniquely understand and enjoy living in terrain normally regarded as inhospitable wilderness. Up until today rearing camels and goats has been the only way in which the bedu can obtain sustenance and cash from the terrain but this does not mean that there will never be other ways in which they might make a living in the Sands in the future. Other ways of making a living, such as tourism, should be looked for. If they can be found and exploited it will be important to the bedu and important to the government which is naturally anxious to discourage too strong a rural-urban drift of the population. Of course, any new exploitation of the region's resources must be ecologically sound. Specialist tourism will be because such tours will not be attracted to the region if it is in ecological decay.

But what is happening to the bedu and their traditional stock-raising activities? The more traditional forms of camel raising will continue, Webster shows (Economy and Society), though altered by the increased use of supplementary feedstuffs. It was not possible independently to assess the impact of these feedstuffs on the total herd size, and the bedu do not think they are increasing stock numbers, but there is evidence to suggest that they are improving the health and condition of camels without increasing their numbers. However, supplementary feeding of goats perhaps has the negative consequence of larger herds of low productivity. This negative trend may also be strengthened by education which may lead, as Webster fears (Gabriel *et al*), towards more part-time and inefficient pastoralism. Smaller but more productive herds should be encouraged but, as Webster (Economy and Society) points out, greater commercial orientation is notoriously difficult since the bedu see herding as outside the commercial domain, enabling them to opt out of or into commercial or wage-earning activities as needed. Nevertheless bedu enterprise is on the increase, largely because of a growing interest in racing camels. Today many bedu in the Sands are, as a part-time or occasional source of income, training or looking after racing camels.

Wells and water availability are a concern of the bedu. Many new wells are being dug, particularly in the *Prosopis* areas (wadis), leading to the practice of some agriculture and therefore to more permanent settlements in the wadis and therefore to overgrazing and to livestock of poor condition. Webster argues that the catchment of rainwater for livestock in cisterns in the central Sands would encourage livestock to stay longer in the Sands after rain where they would make fuller use of the flush of rain-induced vegetation and give rest to the wadis. Munton (Distribution) notes important movements of goats and camels from the woodlands into the central Sands after rain, particularly into the *Panicum turgidum* grasslands. However, stock numbers in the north end of the Eastern *Prosopis* woodland actually increased after the rains and this is clearly associated with overuse and abuse of that area. Change in total stock numbers in the Sands, respectively before and after rain in January/March 1986, was estimated at about 19,000 to 26,000 goats and 3340 to 3193 camels. But although goats can find copious ephemeral forage after rain, Munton (Vegetation and Forage) suggests that because of the *Prosopis* woodlands and the ubiquity of *Calligonum comosum* the Sands afford better all-year browse for camels than for goats.

Farming: Gabriel (Agriculture) states that agriculture in some form employs the vast majority of the people in the Sharqiyah. There are many farming villages at the margins of the Sands, though concentrated in a northerly arc. The older farms are irrigated by *falaj*, with all that that implies for communal organization, but the growing number of new farms lies outside the villages and these farms are well-pump irrigated, giving greater flexibility for organization, volume and frequency of irrigation, cropping, and husbandry and the use of mechanization and expatriate labour. In the old farms dates are the principal crop, in the new farms it is alfalfa, but with a greater concentration on vegetables in the Kamil and Wafi district. There has been an increase in stock numbers in recent years.

Exchanges of goods and services (including labour) between bedu and villager are one more example of the bond that links the central Sands with their margins. These exchanges may be less important than they were because some local goods have been replaced by imports, but they are at any rate undergoing change and the picture is complicated by the fact that bedu own 40 per cent or more of the gardens in some villages.

Fishing: Fisheries, according to Hoek, have the potential to become the dominant source of income for the region. She says that the government intends to develop a large-scale business approach to make optimal use of these rich resources. However, such an approach will require development of a physical, social, and economic infrastructure that at the moment is totally alien to the conservative fishing communities. Many will resent a change from their traditional way of making a living at subsistence level, to compete with the sophisticated fishing industry-to-come. Lancaster (Fishing) would support this contention. He noted a definite antipathy amongst the fishermen and their families to the idea of improving local services, except for the desire for a new clinic, and believes that improved fishing facilities are liked as much because they reduce labour and increase leisure as because they can increase production. But Christie (Regional Fisheries) argues that there is no evidence to suggest that the fishermen between Khuwaymah and Ras Ruways remain subsistence orientated from desire, and she believes that the lack of a good road and of a local ice plant are important factors. Small solar ice plants could answer a real local need. Providing such utilities could, she believes, increase the fish harvest from the villages sufficiently to satisfy both producers and consumers without the danger to the fish resource which could be a consequence of commercialization on too grand a scale.

Crafts: A fairly wide range of crafts is recorded by Crocker including spinning, looping, weaving, sewing, twining, embroidery, palm leaf braiding, and leatherwork. Special regional products include sandsocks and twined rugs whilst camel saddle bags and girth straps are notably well crafted. The market for the camel trappings is probably secure, because of the revival of camel racing noted by Webster (Economy and Society). However, Heath reports that in general the traditional crafts are in decline and she fears that craft skills and appreciation of quality though still present may disappear within the next decade with the loss of the older generation. Already most of the region's village-based craftsmen (potters, pitloom weavers, dyers and silversmiths) have stopped working and the shops are full of cheap imports. Some mothers are still teaching their daughters textile crafts but the demands of conventional education are severely inhibiting this process.

Hoek reminds us that the third five-year plan has envisaged a re-emphasis on traditional economic activities, including crafts. If this idea is to have impact it will require changes to conventional education and training, and the addition of courses in the appropriate technologies. It will also need an intensive input into the promotion and marketing of the goods to ensure that the producers are given the incentive of a satisfactory financial reward for their work. As Heath (Traditional Crafts) puts it: 'Craft skills will revive and continue only if a lively marketing strategy is implemented, and a demand is created.' Equally important is a change in attitude by the craftspeople themselves. Heath noted that the women weavers made much finer camel trappings for their own families than for sale; family values still have priority over financial enterprise.

Women: The future of the crafts, which today are typically in the hands of the women, opens up the question of the future economic role of women in general. There is a danger, as Hoek points out, that development will leave the women with reduced economic opportunities. New jobs in the region are open only to those with at least some education but in the more remote areas, such as Hajj, schooling for girls is not sufficient and parents are reluctant to send their daughters away for education. Shortage of opportunity and cultural norms make it difficult for women, in any case, to enter the job market, and potential women entrepreneurs have problems in gaining access to the banking system. It will indeed be a misfortune if education fails to find a new economic role for women whilst at the same time inhibiting the handing on of their traditional productive craft skills.

Labour migration and remittance wealth: The high level of labour migration from all regions of rural Oman has been noted by many writers, for example Dutton (1983). The migrants typically remit relatively large sums of money back to their home villages. Gabriel (Agriculture) notes the strength of this syndrome in the villages on the Sands margins and reports that out-migration has disrupted the village demographic structure upon which *falaj* organization, the traditional division of labour and the long-established pattern of the agricultural year all depend. There is also a wider wealth gap, date gardens are held in lower esteem, absentee landowners are growing in number and people are 'tied' into the cash economy by the costs of using and servicing their new equipment.

Gabriel (Gabriel *et al*) points out that the earners of remittances are investing it widely, in farms, homes, irrigation machinery, vehicles, consumer goods and small businesses. Lancaster (Gabriel *et al*) shows that although government loans and grants are available to re-equip the artisanal fishing industry the fishermen prefer to use remittance income to buy new capital equipment. Having done so, however, they 'do not seek to maximize the production potential represented by their investment in improved equipment' – as noted above the motivation appears to be as much to do with decreasing labour and increasing leisure as augmenting productivity.

Many men from the bedu families are also migrant labourers. Webster (Economy and Society) notes that pastoralism, never a full-time occupation in the wage-earning sense of the word, gives great flexibility for labour migration and casual entrepreneurial activity.

But Hoek foresees changes. She states that although it is typical, today, for young men to migrate to the Gulf to find work GCC policies may make this less attractive in the future. Already people are returning home because jobs in the Gulf are relatively less remunerative but, with limited education, they are finding it difficult to find work at home.

Interdependence and individualism: As labour migration and remittance money alter traditional patterns of work and wealth creation they are also leading to greater individualism. Lancaster (Fishing) emphasizes the value which the bedu traditionally place on the creation and maintenance of social and economic networks, and the skills with which they use these networks for their advantage. Christie (Markets) illustrates the interdependences which exist between the pastoralists, fishermen and villagers of the region through the processes of marketing and of the exchange of goods and services. The work of all the human scientists, notably Hoek, also shows the great increase in the role played by central government within the local networks. Although the impact of government and of wealth coming from outside the region is demonstrably beneficial in a great variety of ways there is a risk that the resultant reduced dependence on local resources coupled with a shift in responsibility for them from the region to the Capital, a shift noted by Gabriel (Gabriel *et al*), will lead to their mismanagement as local collective responsibility for them gives way to an individualistic tendency to think in terms of the short-term advantage to be gained from their immediate exploitation.

Gabriel notes a trend towards individualism in many spheres of life in Oman, both locally and nationally, a trend epitomized by the move from *falaj*-irrigated gardens to well gardens where decisions can be taken 'relatively unfettered by other members of the community'. Webster (Pastoral Ecology) is concerned about the ecological dangers of increasing individualism resulting in the abandonment of the self-imposed restraints on the exploitation of natural resources, particularly those in the *Prosopis* woodlands.

Government services and private enterprise: The government must control the negative aspects of 'individualism' whilst encouraging the positive aspects of private enterprise. Since 1970 the government has taken on an enormous public sector programme of work, with massive achievements to its credit, but the creation of a dynamic and responsible modern private sector in the Sands region remains elusive. There is a worldwide tendency, as Christie (Gabriel *et al*) points out, to be over-optimistic about the potential of government-controlled systems and to underestimate the potential of the private sector. Christie is talking about marketing but the same dictum could be applied to other sectors of national economies. Oman, with the bulk of its revenue stemming from oil and flowing straight into government coffers, and starting with very few utilities and services in 1970, has concentrated hard on the development of a comprehensive economic and social infrastructure with great speed and effectiveness. Around the Sands today there is a good communications network, and the villages have schools, hospitals and clinics, social services, power stations, development banks and extension centres, and new shopping centres and houses. In the process many new public sector jobs have been created for Omanis, and many expatriates have been drawn into the private sector. Gabriel (Gabriel *et al.*) believes that the drive to install infrastructure has left the private sector relatively neglected, and may indeed have inhibited its growth.

Christie (Markets) points out that the speed of change, induced by new oil wealth, has overwhelmed the ability of the rural sectors effectively to respond. Many production decisions are still based on the subsistence ethic and traditional regional interdependences, and marketing infrastructure is still rudimentary. The rapid growth in the urban demand for food of high quality has, therefore, led to a flood of imports. But the new markets exist, and the local fish traders have demonstrated that the new road networks make the markets accessible as far away as the Gulf coast of Saudi Arabia. Where the fish traders already go, others may follow.

But whilst education and training, coupled with imaginative ideas such as the subsidized apprenticeships advocated by Hoek for young Omanis, may help find employment for people in the private sector, a modern private sector will only grow if traditional entrepreneurial confidence is maintained and bolstered by a combination of appropriate management skills and tempting opportunities to penetrate the modern economy. Encouragement of the private sector needs to be more overt. As Hoek says, whilst the Oman Development Bank, for example, has introduced a loan scheme for small entrepreneurs and the overall government programme of loans and grants is designed to encourage the exploitation of natural resources, these schemes need to be made more widely known within the region. They need to be widely promoted to ensure an attitudinal switch in their favour by potential applicants. They also need to be simple in their operation, and any new small businesses which they stimulate will need effective support (particularly with management skills) in their initial stages.

Tourism: One area of opportunity for the private sector is tourism. As Munton (Evaluation and Management) says, the Sands are of such striking beauty and interest that they are a resource for local and international tourism. Such tourism would have to be, as mentioned above, of a specialist nature and carefully controlled because the Sands are potentially dangerous to casual visitors, and because excessive vehicle use on an ever widening network of tracks destroys vegetation and the animal life which depends on the vegetation. Moreover, insensitive visitors in large numbers will disturb the people and their livestock, and the gazelle for which the Sands are one of the few remaining refuges in Oman.

Nevertheless, tourism gives scope for an imaginative 'Tourist Information Centre' and a tourist base at the north end of the Sands and it will require the services of local guides to conduct visitors through the Sands (possibly on camel back) and explain to them the evolution of the Sand Sea and the fascinating details of its ecosystem. The Centre, the tourist base and the guides will all create forms of local employment which use the unique knowledge of the Sands' people and allow them to combine local money-earning activities with their traditional way of life. Some of the guides can also double as rangers guarding and monitoring the use of the *Prosopis* and other vegetation, and protecting the gazelle and other mammals. A precedent for this is in the Jiddat al Harasis where the White Oryx rangers also look after the gazelle, and they guide parties of visitors whilst maintaining their traditional lifestyle. Tourism will also give a new stimulus to the traditional crafts of the region and thereby answer Heath's call (Traditional Crafts) for new market opportunities. But, if handled sensitively, tourism's greatest value will be to combine new economic activity with positive conservation. Visitors will be coming to see the Sands ecosystem in operation, and if the ecosystem is allowed to fall into decay by its local guardians they will not return.

Communicating the Results

Scientific research papers, now being published in specialist journals, are one means of communicating the results of the Project. Such papers are widely available to the international scientific community. However, these papers, published in a great range of journals, are not necessarily available to, nor written in a suitable form for, scientists and others specifically interested in Oman, nor for Omani decision makers, hence the 'Mapping phase report', the 'Rapid Assessment Document' and this research monograph.

Research information can also be presented in a form suited to the needs of education other than at the scientific research level: written into school textbooks, produced as illustrated booklets and as educational packs. Films and photographs are other media useful for transmitting the research information to wider audiences. Educational visits to projects in the field can be a most stimulating way to communicate the work, particularly to the young – the next generation of scientists. Finally, samples and selected specimens of plants, animals and artifacts can be lodged in museums. Using all these means the Project, from the outset, has aimed to communicate its research findings to as wide an audience as possible.

INTERIM REPORTS

A report on the mapping phase of the Project (Dutton *et al*, 1985), produced by the entire mapping phase team, was published in June 1985 by the Royal Geographical Society and distributed widely within Oman.

A Rapid Assessment Document (Dutton, 1986) was produced by the entire main phase team at a research workshop based in Sultan Qaboos University (SQU) from 1st to 15th April, 1986 immediately following the main phase fieldwork programme. The advantages of producing this report before the team left Oman were various. Field impressions were fresh in people's minds, the entire team was still together so that individual and group discussions could take place freely, and there were no other distractions and responsibilities to attend to. It was also possible to discuss the work with Omani government officials, and other people knowledgeable about the country, during three days of discussion meetings held at SQU. In consequence, it was possible to involve the government more closely in the Project's work and to shape the report to particular Omani government requirements. Copies of the report were delivered, before the main phase team left Oman, to all ministries and other organizations with interests in the Sands.

The report emphasized information of practical relevance for development, for conservation and for education. The term 'practical relevance' was deliberately defined widely to include information leading towards:

* an increase in regional economic activity to the benefit of local people
* the enhancement of people's standard of living
* the formulation of appropriate conservation policies
* the production of local scientific research data for teaching at SQU and school levels
* base-line data for on-going research programmes involving SQU staff and students alike, and
* an understanding of the Sands and the sand sea dynamics in ways relevant to the needs of Omani organizations.

THE SYMPOSIUM AND THE RESEARCH MONOGRAPH

Scientific research papers were produced by all members and associate members of the Project team and presented at a symposium held at SQU in April 1987. Subsequent to the symposium, the papers were revised and new papers appeared as more material was analyzed. All were reviewed by other specialists in their fields and prepared for publication in this research monograph.

RESEARCH PAPERS

In addition to preparing papers for the research monograph, team members, and specialists given responsibility for examining some of the specimens and samples collected in the Sands, will continue to publish scientific papers in specialist journals. The Royal Geographical Society is keeping a central reference library of all these publications which is available on request.

EDUCATIONAL AND POPULAR MATERIAL

A range of other sorts of publications stemming from the Sands Project is beginning to appear. Opportunities for educational and other uses of the Project data are limitless; some are listed below.

Multi-media teaching pack for UK secondary schools: With the assistance of BP Educational Service in London, a multi-media pack for secondary schools (including slides and a map and costing £11.00) has been produced for the General Certificate of Secondary Education Geography examinations (Ridley, 1988).

Multi-media teaching pack for Omani secondary schools: Similar material to the above, but in Arabic, will be used by Oman's Ministry of Education and Youth in secondary schools, published with the help of financial sponsorship by B.P. Arabia Agencies, Oman. This is a very significant addition to locally based educational material on desert development, ecology and change.

Sands booklet: The Ministry of Education and Youth requested that material from the Project should be used for the production of a book for elementary schools. The material has been collected from project members and an Arabic translation of the edited version has been handed over to the Ministry.

Secondary school geography textbook: At the request of the Ministry of Education and Youth, the relevant sections of the secondary school Geography regional and systematic textbooks have been revised by Professor Denys Brunsden and Professor Ron Cooke (consultants to the Project) for inclusion into the next editions.

English language videos: SQU and the British Council in Oman have produced three English language videos based on discussions with the Project directors, each relating to a theme taken from one of the scientific programmes. The videos will be used to introduce new students at SQU to English language scientific discussion of broad ecological interest.

English language reader: An English language reader about the Sands and the findings of the Project has been prepared with the help of independent English language experts working within the Ministry of Social Affairs and Labour. This will contribute to the growing pool of TEFL materials based on Oman, as well as increasing their non-fictional range.

Slide-tape packs: Some 3,000 colour transparencies (slides) were taken by project members and have been selected, catalogued and stored at the

Royal Geographical Society, in London. Some of these slides are being utilized in the publications mentioned above.

Technical and Administrative Report: A report on the technical and administrative aspects of the Project (Holman, 1988) has been published to assist those wanting to undertake future projects of this kind.

PROJECT VISITORS

Students: With the support of the Ministry of Education and Youth ten third-year secondary students were selected from the Mutanaibi Boys School at Ibra to visit the Project once a fortnight. It was not possible to have a mixed group and unfortunately time and resources were not sufficient to run a simultaneous programme of visits for a group of girls. Each visit was given a different scientific emphasis. Members spent time explaining and discussing their work and demonstrating equipment at Taylorbase and in the field, including Fieldbase where several of the research programmes were permanently sited.

The enthusiastic interest of the school-children visitors has two striking implications. Firstly, it will encourage those at SQU whose intention it is to include similar field work and demonstrations into courses. None of the boys had previously travelled into the Sands, despite four of them being from Badiyah with the Sands on their doorstep. The visit to Fieldbase was undoubtedly the highlight of the series, and given less constraints on vehicles, more time would have been spent in the field and less at Taylorbase. The second implication, is that the inclusion of student visits into future projects will be successful. Valuable information was gained from the students because of their local knowledge and contacts, and an improved understanding of the work of the Project by the local community was brought about.

Other school visitors: In addition to the above programme, several other groups of students made day visits to the Project. These included the Sultan's School, the Muscat English Speaking School and the Sumayl Girls Secondary School at Ibra. Each of the groups had quite different objectives for their visits ranging from an emphasis on language to an opportunity to introduce students to the Sands. All of the visits were felt to be equally worthwhile from the point of view of the Project and they heightened the awareness of young people to the diversity and delicate ecology of the Sands.

General Conclusions

INTRODUCTION

It has been a primary intention that the work published in this monograph should be of sound scientific value, and therefore an indispensable reference for anyone undertaking future research work in the Sands, or indeed in other desert areas in Oman. It is also hoped that the multi-disciplinary approach adopted for this Project is justified by the degree of effective and productive communication between the scientists as reflected in their papers here included. However, as the overview in this chapter has tried to demonstrate, the monograph has other values which can be placed under the headings: education, conservation, development and planning. Although it was never an aim of the Project to list a detailed set of proposals for development and conservation, or for research and education, nevertheless several projects have suggested themselves and discussions about them have been initiated with various branches of the government of Oman.

Some follow-up work involving project staff has already started. Gallagher and Hoek, who were already based in Oman, have made several more research visits to the Sands. Christie continues her marketing work in close association with the Public Authority for Marketing Agricultural Produce. Brown is on a two-year contract with the Ministry of Agriculture and Fisheries to study and to plant *Prosopis cineraria* trees. Anderson is working on flood studies with the Council for the Conservation of the Environment and Prevention of Pollution. Warren and a group of geomorphologists have continued research on sand movement and have taught at Sultan Qaboos University. Munton is employed by the Ministry of the Environment and Water Resources with responsibility for the implementation of the Restricted Areas Systems Plan and for the National Conservation Strategy. In both these roles this volume is a principal reference not only for the Sands but also for other desert areas of Oman. Other project staff are also likely to return to Oman as a result of discussions in hand as this volume goes to press.

RESEARCH

The wealth of new information contained in this monograph, and its diversity, confirm our original belief that the Sands are unique and of special research interest on a national and an international scale.

For example, the aeolianite rock, which underlies much of the modern dunes, is unique in extent in the world. The marine margin, though not unique, is an unusual feature of a sand sea, and is of interest as a source for much of the sand, as a rich fishery actively exploited by fishermen 'bedu' living in small settlements along the shore, and as a source of food for a large and varied wader and migrant bird population (particularly along the mudflats bordering the Barr al Hikman).

The research also demonstrates, by the enormous quantity of data that was collected in such a short time from every part of the Sands that this particular sand sea makes an excellent living laboratory, a natural focus for future research on sand sea ecosystems, and a suitable site for a permanent research station. It is a perfect sand sea yet small enough to be comprehended as a whole. It is also isolated and contained but very accessible to properly equipped groups. Base-line data now exist across a wide range of disciplines.

Like all good research programmes the Project has yielded a mixture of answers, partial answers and new questions. Many of the papers suggest new studies that can valuably be undertaken. The Sands could be a focus of scientific research for Sultan Qaboos University, yielding information of value to Oman and to other countries in the Arabian peninsula and elsewhere. Some of this research will add to our broader understanding of the forces which shape the planet, other research will be directed towards questions of more immediate local concern about priorities for conservation and development. All the research will have an important local educational value, particularly if led by staff and students from SQU.

EDUCATION

Conventional, modern education has also come to the Sands. Most children, except apparently some children of the fishing community and of pastoralists living in remote parts of the Sands, attend school. This is admirable but judging from experience elsewhere suggests that these children will develop conventional knowledge at the expense of the understanding of and interest in their local environment that they would otherwise absorb from their parents and relatives. Indeed, some of the fishermen appear to distrust education for this reason, and it has been shown that education is inhibiting the transfer of weaving skills from mothers to daughters.

Such a breakdown in the transfer of indigenous knowledge will foreshadow a lack of respect for local resources. Therefore, as the children of the area are not learning about their environment in the traditional manner because of their modern, conventional education, issues of environmental conservation must be built into the school curricula.

In addition to formal education within a school curriculum there is an opportunity for education in its widest definition; communication to as wide an audience in as many ways as possible. Already, slide-packs, a multi-media teaching pack, a sands booklet, a reader and English language videos are available (see details above), or will soon become available. If they succeed in reaching a wide and youthful audience their impact will be profound, shaping or re-shaping attitudes that will eventually manifest themselves in conservation and development work of sensitivity as the young mature and become the decision makers of tomorrow.

CONSERVATION

The people of the region have evolved economic and social systems in keeping with their environment. They have adopted, through long experience, conservationally sound ways of exploiting local resources – an example being the self-imposed regulations governing the use of *Prosopis* trees.

However, in recent years the outside world has imposed itself on the Sands and its inhabitants much more overtly than before, as a result of the advent of oil wealth. Much of the change is good, of great benefit to the local populations and of little danger to the resources of the area on which the people and their livestock depend. But because, today, much of the private wealth of the area arrives from outside – in the form of remittances – the dependence of the inhabitants on local resources has greatly diminished. This change carries with it the likelihood that their knowledge of and their understanding of the value of these resources is also diminishing with a consequent lessening of interest in using them with traditional skill and wisdom.

The *Prosopis* woodlands on the eastern and western margins of the Sands and in the central Sands play a dominant role in the ecology and in the pastoral economy of the whole area. They check the movement of wind-borne sand, they create a humid micro climate, they provide food and shelter for invertebrates, amphibians, reptiles, birds and mammals, they provide a flight path for birds on migration, and they provide excellent camping grounds for the pastoralists and shade and sustenance for their livestock. In the east, however, they are close to a number of rapidly expanding villages and they also occur over a water resource which is of medium to good quality, which is relatively close to the surface and which is large enough to attract the attention of municipal water suppliers and diesel-pump farmers. If the villagers misuse the trees and if the water resource is over-developed and the water table falls it is likely that the *Prosopis* trees will die and with them the ecology and the pastoral economy of the entire region. Unfortunately there are signs that the traditional respect shown by the people for the trees is beginning to break down, but the Ministry of Agriculture and Fisheries has started a new programme to study and plant *Prosopis* trees, and employs Kevin Brown, who made the study of the trees on the Wahiba Sands Project, to work with Mr Sa'id al Alawi, the Ministry's forestry officer.

Other ecological pressures are mostly at the eastern margin of the Sands where villages and irrigation agriculture are expanding. Wiltshire warns against the indiscriminate use of pesticides. Büttiker and others illustrate the importance of knowing the real insect pests, and their habits and habitats, so that these insects can be controlled without destroying the others.

The Sands themselves remain too hostile to development for much harm to be done, and indeed they provide a refuge for gazelle. The insensitive use of vehicles is probably the main danger. However, the marine margin of the Sands is a special case. It is likely that the track linking the fishing villages will be improved. This will attract casual visitors who could easily upset the waders and migrant birds that feed at the coast, particularly on the mudflats of the Barr al Hikman.

DEVELOPMENT

But the need to conserve local resources must not, of course, be equated with an attempt to freeze all economic change and development. Change, in any case, is occurring – rapidly in some sectors – by a combination of local initiative (investing remittances) and government support. One example of government support is the operation of recently introduced systems of modern financing by the development banks. An example of entrepreneurial development is in the marketing of fish, often as far afield as the Gulf states. Inevitably there are bottle-necks in the development process, and we have tried to define where these exist and how they might be remedied.

However, the Project has not identified many new or untapped resources of great economic significance. The exception may lie in the beauty, the fascination and the challenge of the Sands themselves and in their consequent attraction to visitors. There is, therefore, a visitor resource, discussed by Munton (Evaluation and Management), which could bring people from all walks of life into the Sands: school children, scouts, students, internal visitors and, eventually, foreign tourists. Properly managed such an enterprise could bring: knowledge of Oman to a widening international audience, greater understanding of life beyond the towns to a new generation of urban children, protection to the local wildlife and employment to local people.

PLANNING

The achievement of an appropriate balance between enterprising development and conservation in the Sands will depend on an accord between all interested parties, who will need to work together towards the implementation of an agreed Plan. It looks as though this may come about. Munton (Evaluation and Management) reminds us that Oman is giving full consideration to a Proposal for a System of Nature Conservation Areas. In this scheme the Sands will be designated a National Resource Reserve (NRR) indicating that it is worthy of special study pending subsequent classification as either a National Nature Reserve (NNR) or a National Scenic Reserve (NSR). The latter category, because of permanent settlement or lesser value than the former, will be less closely managed.

Munton argues that as far as the Sands (including it is hoped, part of the raised channel area to their west) are concerned the primary management objectives are: to preserve the diversity of the dunes and the raised channels; to maintain the populations of *Prosopis* trees and other plants and of the gazelle and other animals; to conserve and enhance the rangeland resource; and to protect the relic *Prosopis* woodland for its economic and wildlife importance.

However, for a Plan for the Sands to be successful the bedu will have to be active partners in both its design and implementation. Webster (Pastoral Ecology) notes that the increasingly intensive use of the forage resources and the trend towards settlement by the bedu are having detrimental consequences, particularly in the *Prosopis* zones, whilst the traditional means of preventing overexploitation of these areas are weakening. But their revival, he suggests, may yet be encouraged as part of a conservation and management plan for the region if the wellbeing of the bedu livestock is fully taken into consideration. Webster believes that the bedu would then respond positively to initiatives such as tree planting and the controlled cultivation of forage crops. Such programmes by themselves, however, will tend to encourage more settlement and therefore more strain on limited areas of the total Sands forage resource. Any Plan must include schemes to retain the traditional seasonal mobility of the pastoralists in order to spread resource usage as evenly as possible. Webster mentions: strengthening and extending the traditional practices of making temporary reserves and of the ban on tree felling; encouraging the use of building materials and fuels other than *Prosopis* wood; and improving rainwater storage in the central Sands, as mentioned above.

Webster also points out that a Plan will, if the bedu are brought into the planning procedure at an early stage, answer the special needs that the bedu have for utilities and services that are so designed and located (or even made mobile) that the bedu can maintain their preferred lifestyle and remain as 'guardians' of the Sands whilst obtaining their fair share of the amenities that the government is providing for everyone in Oman. In conclusion, it can be said that the Sands and their margins are an area of great natural beauty, of equal scientific interest and ecological value, and the resource upon which their inhabitants depend. They are a very special part of Oman's heritage. Long may they be treated with the interest and respect they deserve by both local people and visitors alike. The Wahiba Sands team hope that this monograph will provide a research platform on which can be built integrated research, conservation and development programmes of enduring benefit under the aegis of Oman's Ministries and University following the principle of 'shared responsibility' given in the foreword to this volume by His Majesty Sultan Qaboos.

1 This paper first appeared in *The Journal of Oman Studies* Special Report No. 3, 1988, 'The Scientific Results of the Royal Geographical Society's Oman Wahiba Sands Project 1985–1987', published by the Office of the Adviser for Conservation of the Environment, Diwan of Royal Court, The Palace, Muscat, Oman.

The contents of this Special Report include:

Bedouin Ecology and Management	*R Webster*
Distribution of Livestock	*P Munton*
Bedouin Economy and Society	*R Webster*
Agriculture	*T Gabriel*
Fishing Communities	*W Lancaster*
Fishing Economy	*A Christie*
Traditional Crafts	*G Crocker (with C Heath)*
Markets and Marketing	*A Christie*
Government and Economic Growth	*C Hoek*
Government Incentives	*T Gabriel et al.*
Evaluation and Management	*P Munton*

Acknowledgements

The Author, on behalf of the Royal Geographical Society wishes to acknowledge the very wide interest shown in the work of the Project and the generous support it has received from so many individuals and organizations both in the Sultanate of Oman and the United Kingdom over the past four years. Team members have been overwhelmed by this 'orchestra' of support and would like to convey their heartfelt thanks and appreciation. Such diversity of committed support has enabled the Society to undertake this detailed investigation of the Sands and collect considerable data for Special Report No. 3 of the *Journal of Oman Studies*, edited by Dr Roderic Dutton.

The following are responsible and deserve the fullest credit possible:

BY KIND PERMISSION OF HIS MAJESTY SULTAN QABOOS BIN SAID

SPONSORED BY THE DIWAN OF ROYAL COURT AFFAIRS
President: His Excellency Sayyid Saif Bin Hamad

PATRON: HIS ROYAL HIGHNESS PRINCE MICHAEL OF KENT

The eight CORPORATE PATRONS for underwriting the major costs of the project.

ASSARAIN ENTERPRISE (OMAN)
GULF AIR (BAHRAIN)
LAND ROVER LTD (UK)
MOHSIN HAIDER DARWISH (OMAN)
RACAL ELECTRONICS (UK)
SUHAIL AND SAUD BAHWAN (OMAN)
TAYLOR WOODROW-TOWELL (OMAN)
ZUBAIR ENTERPRISES (OMAN)

Brigadier J T W Landon

The SULTAN OF OMAN'S ARMED FORCES
for providing a military solution to a civilian scientific problem.

Very special thanks go to the SULTAN OF OMAN'S ARMED FORCES for providing full logistical and support facilities to the project, enabling some 30 scientists to operate for four months with confidence over a 15,000 square kilometre area of desert sand sea in complete safety and without mishap. Particular thanks go to the three Service Commanders who authorized this support:

Liwaa Nasib bin Hamad bin Salem Al Ruwaihi WO, WKhm, ON(S)	CSOLF
Liwaa (Tayar) Erik P Bennett WO CB	CSOAF
Liwaa (Bahry) Hugh M Balfour LVO	CSON

Members of the Project wish to acknowledge the very special 'pillars' of support they have received from the following individuals:

Abdulla bin Saleh bin Baabood (Assarain Enterprise); Shaikh Amor bin Ali bin Ameir (Vice Chancellor – Sultan Qaboos University); Raa'id and Mrs Chris Beal; John E Cox OBE (Taylor Woodrow Towell); Mr and Mrs Richard Dalton (British Embassy); Ralph Daly OBE (Conservation Adviser, Diwan of Royal Court); Raa'id Bill Davies (Liaison Officer – HQ SOLF); Dr Carroll Hess (Sultan Qaboos University); His Excellency Hussain bin Mohammed bin Ali (Omani Ambassador in London); William W Doyel (Senior Adviser, PAWR, later Council for Conservation of Environment and Water Resources); Aqeed Khamis Bin Mohammad Al-Amry (Chairman of the Oman Co-ordinating Committee); Raa'id (Rukn) Khalfan bin Muzzein bin Baroot; Shaikh Mohammad bin Humaid al Hajri (Hewiyah); Shaikh Muhammad bin Hamad bin Duqhmal Al Wahibi; Mohammed Al Medfa (Gulf Air); Mohammed Saud Bahwan (Suhail and Saud Bahwan); Kamal Abdurrehda Sultan (WJ Towell); Raa'id and Mrs Eileen Scarff, His Excellency Abdullah Sakhr Al Amri (Under Secretary, Ministry of Information); Dr Omar Zawawi (Chairman of Oman International Bank and representing all the Banks in Oman who generously contributed to the Project); His Excellency Mohammed bin Zubair bin Ali.

OMAN CO-ORDINATING COMMITTEE

Chairman: Aqeed Khamis bin Mohammed Al-Amry

Ministry of Agriculture and Fisheries (Dr Abdul Munim M Mjeni); Ministry of Education and Youth Affairs (Khalil bin Hamdan Tabsh); Ministry of Electricity and Water (H E Khalfan bin Nasser al Wahaibi); Ministry of Environment and Water Resources (Dr N J Kapadia, Mrs Jocelyn Kharusi); Ministry of Information (H E Salem Bin Said Al Siyabi, Mohammed Al Mahoon); Ministry of National Heritage and Culture (Sabah Abdul Amir Ali, Dr Paulo M Costa, Michael Gallagher); Ministry of Petroleum and Minerals (Khalifa bin Mubarak bin Ali al-Hinai); Ministry of Posts Telegraphs and Telephones (Ibrahim bin Awad bin Sebyt al-Hassawi); Ministry of Social Affairs and Labour (Mohamad bin Mohamad bin Shadad); Ministry of the Interior (H E Shaik Ahmed bin Saif Al Mahrouki); National Survey Authority (Aqeed Bill Codd); Council for Conservation of Environment and Water Resources (H H Sayyid Barghash bin Ghalib Al Said, Mr W Donald Davison Jr); Royal Oman Police (Muqaddam Abdul Aziz bin Awad Al Ruwas); Sultan of Oman's Air Force (Aqeed Rukn Tayyar Mohamad Bin Mubarak Bin Mahroon Al Amri); Sultan Qaboos University (Dr Carroll Hess).

ROYAL GEOGRAPHICAL SOCIETY OMAN COMMITTEE

Chairman: Colonel Stephen Gilbert

Dr Colin Bertram, Professor Denys Brunsden (Kings College), Professor Ron Cooke (University College), Dr Roderic Dutton (Durham), Sir Vivian Fuchs, John Grannan (Durham – Computing Director), Professor Andrew Goudie (Oxford), David Hall Esq, Dr John Hemmings (RGS Director), Louise Henson (Secretariat), Sir Lawrence Kirwan, Dr Paul Munton (York – Biological Resources Director), Rebecca Ridley (Education and Secretariat), Nick Theakston, (Field Director – Peat Marwick McLintock), Dr Andrew Warren (University College – Geomorphology Director), Dr Roger Webster (Exeter – Human Programme Director), Shane Winser (RGS Home Agent), Nigel de N. Winser (RGS Project Director).

The PROJECT GUIDES, FIELD ADVISERS AND INTERPRETERS

Abdulhakim bin Amir bin Nasser Al Yahamadi (from Shariq)
Ali bin Abdullah bin Ha'atrush Al Wahibi (from Raqah)

Khalifa bin Dhuwayin bin Hilays Al Wahibi (from Mintirib)
Said bin Jabir bin Hilays Al Wahibi (from Mintirib)
Salim bin Hamed bin Rashid Al Hagari (from Mintirib)
Shaikh Sultan bin Nasser bin Mansur Al Ghufayli (from Aflaj)
Shaikh Mubarak bin Hilays Al Wahibi

The Project comprised eight sections and each wishes to acknowledge particular support and assistance as follows:

A. ADMINISTRATION AND SECRETARIAT
B. BIOLOGICAL RESOURCES
C. COMPUTING RESOURCES
D. EDUCATION
E. GEOMORPHOLOGY
F. HUMAN PROGRAMME
G. OPERATIONS
H. SURVEY

A. ADMINISTRATION and SECRETARIAT

Advisers
Ministry of Defence (Mr Peter G Boxhall); Oman Holdings International LLC (Mr Maurice Hynett); Council for Conservation of Environment and Water Resources (Mr James R (Digger) Jones); Suhail and Saud Bahwan (Mr Alex Borges; Mr J P Antia); Sultan Qaboos University (H E Shaikh Amor bin Ali Bin Ameir; Professor and Mrs Jack Hunt, Dr and Mrs Carroll Hess); Sultan of Oman's Land Forces (Aqeed Clive Brennan, Raa'id Chris Benthall-Warner; Raa'id Tim M Cheater; Raa'id Bill Davies; Naqeeb Nigel Sharpe); W J Towell and Co (LLC) (Mr Michael J Beardwood); Mr and Mrs C Weston-Baker; Mrs Mary Sales

Exchequer (Financial Sponsors)
Addison Wheeler Fund; Al Bank Al Ahli Al Omani (Sheikh Zaher al Harthy); Assarain Enterprise (Abdulla bin Saleh Baabood; Mr W J Mintowt-Czyz); Assarain Group (Nanda Shiv Prakash); Balfour Beatty Ltd (Mr Neil Ashley; Mr John Craig); Bank of Oman and the Gulf (Abdulaziz Ali Al Owais); Bank of Oman, Bahrain and Kuwait (Mohsin Haider Darwish; Mr Ian A Marnoch); British Aerospace (Mr Jeremy J. Lee; Ms Anne Dixson); British Bank of the Middle East (Mr H R Williams); Mr Peter F H Mason); Cementation International Ltd (Mr Mike Slater); Commercial Bank of Oman (Maqbool bin Hameed); Duke of Edinburgh's Trust No 2 (Mr V G Jewell); George Wimpey plc (Mr C J Chetwood); Gray, MacKenzie and Co Ltd (Oman United Agencies LLC) (Sir Rae McKaig; Mr Richard Owens); John Laing International (Mr L J Holliday); Laing Oman LLC (Mr Keith E. Blow); National Bank of Oman (Arif Maqsood; Haji Ali bin Sultan); Oman Arab Bank (HE Sayyid Salem bin Nasr al Busaidi); Oman Holdings International (Mr Maurice Hynett); Oman International Bank (Dr Omar Zawawi; Ms Susan Lyddon); Oman Overseas Trust Bank (HE Sheik Mohamed bin Salman); Oman United Agencies LLC (Mr J Thorne); Peat Marwick Mitchell; Royal Oman Police (Inspector Yousef Dadallah Mohd Al Zidjali); Sir Alexander Gibb and Partners (Mr G H Coates; Mr G R Martin); Standard Chartered Bank (Mr A R Holden; Rt Hon Lord Barber); Suhail and Saud Bahwan (Mohammed Saud Bahwan; Mr Antia; Union Bank of Oman (Khamis bin Ali Al Hashar); Widnell and Trollope and Partners (Mr Shawki Sultan; Mr Ian Robertson); Wimpey – Alawi LLC (Mr Hugh Hayden); Mrs Ruth Bennett-Jones; Brigadier J T W Landon; Mr Frank F Steele; Zubair Enterprises (Mr Mohammed bin Zubair bin Ali, Mr M G F Dickens, Mr Anders Ulegard).

Flights
Gulf Air (Hamad A R Al Medfa; Ali Abdulkhaliq; Averil Bloy Slade; Salim Al Salmi; Mohamed Hassan A Rasool; Mohammed Salim Ahmed; Mike Simon; Victor Grant and Gulf Air hostesses everywhere).

Insurance
Muscat Insurance Services Ltd (Mr Robin Fenton; Mr Charles Morris); Oman National Insurance Co (Mike Adams); Sedgwick International Ltd (Mr Tony Wood).

Other Key Contacts
Al – Yousef International Enterprises (Mr Alan R. Milne); British Embassy (His Excellency Duncan Slater; Richard Dalton; Mr Hamilton Marcelin; Mr Roger French; Ms Marie O'Conner); British

Embassy Kuwait (Andrew and ;Christina Heath); Financial Times (Mr Rod Oram); Historial Association of Oman (Mrs Jean Hirst); Ministry of Defence (Aqueed F Tas Fisher; Said Al Riyami); Ministry of Education and Youth Affairs (Soud M al-Timami); Ministry of Electricity and Water (Mustafa bin Mukhtar bin Ali); Oman Aviation Services Co (SAO) (Mr E J P Robinson); Oman United Agencies LLC (Mr Stephen Moran); Peat Marwick Mitchell and Co (Mr Zafer H Siddiqi); Shankland Cox International Ltd (Oman) (Mr Rodney Hamilton); Shell Markets (Middle East) Ltd (Mr David Gillon); Sultan Qaboos University (Prof. Gill Heseltine; Mr Alfred Rigby; Mr Vaughan Volungis; Professor George Gamlen); Sultan of Oman's Air Force (Muqaddam Mohamed Mubarek; Naqeeb (Tayyar) R (Dick) Cowley); Sultan of Oman's Armed Forces (Raa'id Mike Tily); Sultan of Oman's Land Forces (Ameed Khamis bin Humaid bin Hamed Al Kilbani WKhM; Ameed Yussuf bin Khalfan bin Zahir Al Busaidi; George Mathers; Muqaddam Andrew W Swindale; Muqaddam George A Mathew; Muqaddam John A Williams; Muqaddam Saleh bin Ali bin Said Al Yahyai); Naqeeb Keith Day; Raa'id Allan Malcolm; Raa'id Bill H Foxton; Raa'id David McAllister; Raa'id George F Correa; Raa'id Imam Bux; Raa'id Mike Spencer; Raa'id Mohammed bin Ali bin Siddique Al Zidgali; Raa'id Mohammed bin Salem bin Ali; Raa'id R (Bob) J McCartney; Raa'id Said bin Khalfan bin Hamood Al Weidi; Raa'id Simon R Yates; Raa'id Stan D Richards); Sultan of Oman's Navy (Raa'id (Bahry) Alec Tilley); Sultan's Armed Forces (Commander Philip Cookson); United Nations Environment Programme (Mr Michael Gwynne); University Club (Mr Mohammed Gamel); University of Durham (Professor John Clarke); Mrs Mo Codd; Captain Nigel Sharp; Chester and Lucy Williams; Mohammed Ali Al Riyami; Mr Patrick Miller.

Press and Public Relations
Embassy of the Sultanate of Oman (Mr Kenneth Brazier); Geographical Magazine (Mr Iain Bain); Muscat Inter Continental Hotal (Mr Martin Weber; Mr Jonathan Badman; Ms Ann M Styles); Middle East Times (Mr Tim Owen); Ministry of Information (HE Salem bin Said Al Siyabi, Omar and the Hollywoods); Oman Daily Observer (Mr G Reid-Anderson; Mrs Meredith Campbell); Times of Oman (Mr Arthur C Cushing).

Secretariat
British Embassy (Marie O'Connor, Ms Deborah Ford, Kate Livingstone); Alison Edwards; Home Agents Diana Hancock and Amelia Dalton; Ministry of Defence (Mr Arthur Davis MIPS); Municipal Journal Ltd (Ms Pat Reviere; Mrs Louisa Service; Dr John Hemming); National Printers (Mr Iflikar A Malik); Peat Marwick McLintock and Co Ltd (Mr C T E Hayward); All Royal Geographical Society Staff (particularly the Bursar, Fay, and Margaret Sullivan on the switchboard, Ted Hatch, the Map Room, Judith Thomsen and the General Office, Jim and Margaret, the Directors Office and the Library, Prue and Peter), Roneo-Alcatel (Mr J Balson); Sultan Qaboos University (Mr and Mrs Carroll Hess, Mr and Mrs Jack Hunt, Mr Michael Brookes; Ms Carrol Jones; Ms Martha L P Dukas); Sultan of Oman's Land Forces (Raa'id Bill Davies, Nasser bin Ali bin Abdullah; Raa'id Dai Davies); Lucy Goelet; Julia Collomb; Peat Marwick McLintock Will Peskett).

B. BIOLOGICAL RESOURCES

Al Kamil tree nursery; Captive Breeding Centre (Mr Jeremy Usher-Smith; Mr Richard Wood); Basel Natural History Museum (Dr Michel Brancucci); British Museum (Natural History) (Dr P M Mordan; Dr E N Arnold); City Museum of Leeds (C A Collingwood); Flora and Fauna Preservation Society (Mr John Burton); GRM International Pty Ltd (Dr David Hall); Ibra District Hospital (Dr P B Pamar); International Union for the Conservation of Nature (Dr John E Clarke); International Union for the Conservation of Nature: Conservation Monitoring Centre (Dr Mark Collins); Ministry of Agriculture and Fisheries (Kamal Eldin Mohamed Ali MV; Dr Mansoor; Dr R M Ray Lawton; Said bin Hamed al-Alawi); Oman Natural History Museum, Ministry of Natural Heritage and Culture (Sayyid Faisal bin Ali Al-Said; Michael Gallagher); Philip Harris International Ltd (Mr B J F Haller, Mr and Mrs Richard Cochrane; Mr Foley; Mr Stewart Parkin); Planning Committee for Development and Environment in the Southern Region (Dr Robert Whitcombe); Royal Air Force Ornithological Society (Major David Counsell); Royal Botanic Gardens (Mr Gren Lucas, Dr G Wickens – Kew; Mr Tony Miller, Miss Rose King – Edinburgh); Sultan Qaboos University (Mr Lemoyne Hogan; Prof M J Delany); Tropical Development and Research Institute (G B Popov MBE); University of Cairo (Prof Mabil Hadidi); University College London (Professor George Stewart); University of Durham (Dr Phil Gates); University of Kuwait (Prof Laitfy Boulos); University of Kent (Dr Ian R Swingland); University of Oxford (Dr Malcolm Coe); University of Strathclyde (Mr Peter G Waterman); Whitely Animal Protection Trust (Mr J B Dodd); World Wildlife Fund International (Dr Hartmut Jungius, Dr John Hanks); White Oryx Project (Dr and Mrs Mark and

Karen Stanley Price); Harrison Zoological Museum (Dr David L Harrison); Dr Torben B Larsen; Dr Josephine Turquet; Mrs Sue Waters.

C. COMPUTING RESOURCES

The following are to be specially thanked for the design and function of the IBM Wahiba Field Computer System.

3M United Kingdom plc (Mr Nigel Murphy; Mr Tony Hesforth); Husky Computers (Mr Ken Rainsforth; Mr Malcolm Garrett); IBM (Mr Chris Bowers; Mr Max Hoffman; Mr L Lippi; Ms Fiona Worsley; Mr John Latimer); Innovative Software (UK), Inc. (Ms Deborah Whittick; Mr Thomas DeBacco); Qubie (Mr Roger Harvey; Mr Peter Dignam); Suhail and Saud Bahwan (Mr G V Purushotham Rao); Sultan Qaboos University (Dr Ron Cotterall, Barry Rawlins).

D. EDUCATION

BP Educational Service (Mr David M Barnett; Mr Tim Morris); BP Arabian Agencies Ltd (Mr Glen Lawes); British Council (Keith and Gillian Jones; Mr Tony Andrews); Sumaya Secondary School, Ibra (Mrs Mary Griffiths); Ministry of Education and Youth Affairs (Khalil bin Hamdan Tabash; Osman Hassan Abu Darag; Abu Obeida Mohd Ali; Ali Muhsein Al Hafiez; Ahmed Al Zubaidi; Saleh bin Abdullah Al Ghalani; Mohammed Ahmed Omar); Muscat English Speaking School (Ms Julie Mills); Sultan of Oman's Air Force (Hilal Amour); Sultan's School, Muscat (Mr John Chalfont); Sultan Qaboos University (Mr Mike Johnston; Mr Charles Backhouse; Mr Jim Melia; Mr Mike Neff; Mr Vaughn Volungius); RGS Education Committee (Professor W V G Balchin (Chairman) and all committee members); Northgate Training (Mike Lynch); Operation Raleigh (Chal Chute); Bath University (Mr Jerry Hones); Cardiff University (Dr John Yockney); University College (Prof Ron Cooke); Secondary Examinations Council (Mrs Pat Wilson); Ibra Al Mutanabi Secondary School (Abdullah Salim Rashid al Hajari; Saleh Juma Al Blushi; Mohemed Hamdan Al Hajari; Mohammad Hamood Al Blushi; Masoud Saloum Al Shoukairi; Salim Abdulla Hamed Al Jufally; Hamoud Nasser Al Sinawi; Hamad Al Sinawi; Abdul-aziz Salim; Obeid Said Al Shukiry).

E. GEOMORPHOLOGY

Al Ghaina Hospital (Gurbinder Singh; Dr Said Arsal Ahmed); Al Mansoori Oil Camp (Mr Ken Martin); Amoco Oman Oil Co (Dr Alison Ries; Mr Chuck Pitman; Mr Max W Leenhouts); Apex Publishing (Mr Andy Mitchell); Camerino University (Mr Mauro Coltorti); Feslente Fordath (Mr G Bamford); Hayter Fund (Mrs Phillida Dann); Japanese Petroleum Development Co (Oman) Ltd (Mr Fukushima; Mr Oguesa; Mr Tusuda; Mr Uchiyama); London School of Economics (Mr Gavin Allan-Wood; Mrs Jane Pugh); London University Central Research Fund, London University Hayter Fund; Ministry of Communications (Abdul Latif Huneidi; Ahmed Hamood Al-Harthy; Mohammed bin Salem Al-Bussaidi; Mr Nevill Nisbet; Mr Peter Smith; Mr Maurice Lane); Ministry of Health (Mohammed Bakr bin Moosa bin Ali); Ministry of Petroleum and Minerals (A K Abu Risheh; Abdul Hussain bin Ahmed Al-Eisa; Mohammed bin Hussain bin Kassim; Mr A P (Sandy) Leggat; Mr D V Parker); Ministry of Agriculture (Dr Ramzan); Natural Environment Research Council UK (Dr Steve Donovan; Dr A Thomas; Ms Mary Coole); North East London Polytechnic (Dr Bruno Canzini); Pacific Consultants International (Mr Hisao Wushiki); Pan Am (Mr Lyn Batstone); Petroleum Development Oman (Dr C H Mercanton; Dr Michael Hughes-Clarke; Nasser Abdullah Al-Lamki; Mr C Pryce; Mr R Kneijsborg; Mr P Duff); Public Authority for Water Resources now – Council for Conservation of Environment and Water Resources (Mr William W Doyel; Mr Robert J. Dingman; Don Davison; Graham Munroe-Thompson; J R 'Digger' Jones; Mr Solly Purian; Ms Harriet Nash; Patrick Considine; Rod and Margaret Mitchell; Mr John Kay; Mr Bob Koepp; Mr Gordon Stanger; Mr Horst Weier; Mr Chase Tibbitts; Mr Mark Tomlinson; Yasser bin Salim al Harthy); Mul/2 Said bin Mubarik bin Saleh Al Bahloli and the Coast Security Force soldiers at Field Base; Shell Exploration and Production (Dr Ken Glennie); Sultan of Oman's Navy (Lt Cdr Peter Clover; Raa'id (Bahry) Peter Banyard; Bruce Thompson); United Nations Environment Programme GRID (Dr D Wayne Mooneyhan); University College (Dr Claudio Vita-Finzi); University of Durham (Dr Ray Harris, Dr Tim Munday); University of Oxford (Mr Chris Jackson); Mr Tony Wilkinson; Mrs Judy Haines; Mrs Kathie Smythe; Maj Tim Cheater and Capt Bill Rudd of the Desert Regiment and Sgt Sulayam.

F. HUMAN PROGRAMME

Agricultural Research & Development Project, UNDP (Mr Michael Hyland; Dr Ramzan Muhammad); Bank of Agriculture and Fisheries (Ahmed bin Mohammed Al Massan); Development Bank (Amin Baker); Frederick Soddy Trust; GRM International Pty Ltd (Dr Russell Riek; Dr Stewart Routledge); Japan Petroleum Development Oman (Mr Kenichi Fukushima); Ministry of Agriculture and Fisheries (Dr Muhammed Ramzan; Rashid Amor al Barwani; Mohammed Redha Hassan; Hilal Zaher al Kindry; Ahnef Zubaidi; Ali Amri; All staff at Kamil-Wafi HQ; Mr Michael Kerr; Fonad Yehia; Ibrahim Salah); Ministry of Commerce and Industry (Hilal Kiyuni; Hamza A Al Asfoor; Khamiz Al Khyomi; Salah M Abu); Ministry of Education and Youth Affairs (Sadik Yaffer Mohammed); Ministry of Social Affairs and Labour (Salim Reamy); Oman Flour Mills Co Ltd (Mr Donald A Macdonald; Mr Jan F Watson); Public Authority for Marketing Agricultural Produce (HE Said Nasser al Khusaibi; Suleiman Amer Al Mahrazi; Abdul Aziz Al Areimi; Adel M Salem Khalil; Nasser M N al Hadhrami; Mr John Priest; Mr Jim Moore; Humaid Salim al Alawi); Resource Development Associates (Mr Scott McEntire; Mr Tony Rasch; Mr Stanley Swerdloff); Sultan Qaboos University (Prof G F D Heseltine; Dr Carroll Hess; Prof Clive Holes; Mr P Jim Melia); Toyota, Ibra (Hamed Mubarak Al Mudalwy); Sultan's Armed Forces (Ms Angela Roddis; Raa'id Tim Cheater, Naqeeb Bill Rudd – Desert Regiment); Union Bank (Nasser M Harassy; Said A Mahrooqi); Vocational Training Institute, Ibra (Mr Issa Rashdy); W D Scott, International Management Consultants (Mr Richard Britton); Professor Dale F Eickelman; Abdul Hakim Amor; Dr Miranda Morris; Mr Masoud Nasser Al Sheibany; Mohammed bin Shamis Al Wahibi; Mohammed bin Ali Al Amiri; Dr Dawn Chatty; Abdullah Muhammad Al Salimi; Ali bin Hamad bin Lahush al Wahibi; Ghudayyir bin Sultan al Wahibi; Hamad bin Dhuway'in al Yahhafi; Muhammad bin Salim al Wahibi; Muhammad bin Hamad al Wahibi; Latifa bint Jabir bin Hilays al Wahibi; Hadoob bin Sangoor al Hajari.

G. OPERATIONS

Sultan of Oman's Land Forces (Raa'id Khalfan bin Muzzein bin Barrot Al Woshahi; Naqeeb Ali bin Hamood bin Humaid al Fara'ai; Naqeeb C L Griffiths; Wkl/2 Sobait bin Mubarik bin Khamis Al Alawi; Rqb/1 Mohammed Ashraf; Rqb Saleh bin Barkat bin Yahya Abdiali; Naqeeb John Wiegold; Raa'id Mike Lobb; Raaid Tim Cheater (DR); Naqeet Bill Rudd (DR)).

Taylorbase, Capital 'Heide' Base and Field Bases
Khimji Ramdas (Ajay M Khimji); Ministry of Defence Engineering Division (Mohammed Hamdan Nasser Al Sawafy; Mr Andy A A Allan; Mr David Markham; Mr Dick D Holland; Mr Frank Gaughan; Mr Greg M Raven; Mr J Mike Holliday; Mr Kieth Bull; Mr Mike J Fordham; Mr Mike S Carter; Mr Sid Rowe); Sultan of Oman's Land Forces (particularly the Coast Security Force and the Desert Regiment (Muqaddam Salim bin Jumma bin Aboud Al Shaqsi; Raa'd (Rukn) Khalfan bin Muhsin bin Baroot Al Woshahi; Naqeeb Ali bin Hamood bin Humaid Al Fara'ai; Naqeeb C L Griffiths; Muqaddam Abdullah bin Amor bin Ali Al Harthi; Raa'id Harry Boseley; MUL/2 Said bin Mubarak bin Saleh Al Bahloli; MUL/2 Khadeem bin Saif bin Khadeem Al Muqbali; MUL/2 Jumma bin Ibrahim bin Khamis Al Baluchi; MUL/2 Hilal bin Said bin Hamood Al Hossani; Rqb/1 Salim bin Ali Khamis bin Saif Al Alawi; Rqb/1 Salim bin Ali bin Nassir Al Harthi; Rqb/1 Ghasib bin Hamed bin Saud Al Siyabi; Rqb/1 Hamed bin Salem bin Said Al Habsi; Arf Mohd bin Humaid bin Mohd Al Awaisi); Sultan's Armed Forces Engineers (Arf Amor bin Rashid bin Amor; Muqaddam Hamed Said Mahruki; Muqaddam Mohammad bin Sheikhan bin Zahar Al Kindi; N/Arf Mohammad bin Suleman bin Salim; Naqeeb Jim W Mitchell; Raqeeb Ian Sinclair RE); TAYLOR WOODROW-TOWELL (John E Cox OBE; Mr and Mrs Terry Tobin, Mr Andrew Leslie, Mr Takla H Takla; Mr Alan Baldwin; Mr Archie Johnstone; Mr Harry Melling; Mr Ken Cain; Mr Laurian Perera; Mr Robert Warmwell; Mr Singh; Mr Tony Hyde – for building us the best Base Camp in the world and Rex Cox).

Catering
GETCO Group (Mr G L Johnson); Matrah Cold Stores (Mr Tony Perks; Mr Valsan); Muscat Inter Continental (Mr Jonathan Badman; Ms Ann M Styles); Party Ingredients Ltd (Mr Peter Gladwin); Spinneys (1948) Ltd (Mr Verity); Sultan of Oman's Land Forces (Mr John Augustus and the CSF Mess Staff; Mr Paul Silveira; Muqaddam Nick A Alister Jones).

Medical
Armour Pharmaceutical Co Ltd (Mr P Gallard); Astra Pharmaceuticals Ltd (Mr Stuart C Elder); Bayer UK Ltd (Ms C Bargeron); Beecham Research Laboratories (Mr W M Burns); Duncan,

Flockhart and Co Ltd (Mrs P M Simms); E R Squibb and Sons Ltd (Mr K A J Ratcliffe); Fairey Industrial Ceramics Ltd (Mrs Maureen Perkins); ICI plc (Mr T J Veal); Ibra District Hospital and Dental Dept (Dr Jyentha; Mrs Freda Barlow; Salim Amor Al Harthy); Kirby-Warwick Pharmaceuticals Ltd (Mrs O M Bovill); Lakeland Plastics (Mr S A M Rayner); May and Baker Ltd (Mr I P Arnold); Medident Centre (Dr Rosemary A Leila); Ministry of Health (Mohammed Baker bin Musa bin Ali); Mintirib Hospital (Dr Kamal Mohammed); Napp Laboratories Ltd (Ms Genita Whittingham); Reckitt and Colman (Overseas) Ltd; Roche Products Ltd (Mrs J A Kilshaw); Rocket of London Ltd (Mr Bradley); Searle Pharmaceuticals (Mrs A E VasBenter); Smith and Nephew Ltd (Mr W. Dixon-Payne); Sultan of Oman's Land Forces (Aqeed L C Banks OBE ME BCh BAO MFCM; Muqaddam Lutfia bint Ali bin Said Al Kharusi; Rqb/1 Said bin Khalfan bin Said Al Mandhari; Naqeeb Ruddy (DR)); Sultan's Armed Forces (Aqeed H Lomax Roberts MB ChB DTM&H MBIM; Ms Angela Roddis); The Radcliffe Infirmary (Dr Bent Juel-Jensen); Travenol Laboratories Ltd (Mr J D Nicholls); Utila Geratebau GmbH + Co KG (Herr Geyr); Warner-Lambert Central Africa (Mr D J McClelland); Wellcome Foundation Ltd (Ms Anne Marsh); Windsor Pharmaceuticals Ltd (Miss Griffiths); Winthrop Laboratories (Miss Louise Ducker).

Photography
Barker, Barton and Lawson (Oman) (Mr Simon Ferrey); Camera Care Systems (Mr Nigel Gifford); ECO Communications (Dr Thomas Schultze-Westrum); Fuji UK (Hanimex) (Graham Rutherford; Hailey Beale; Dave McCubbin, Rebecca Brain; Anne Mapson); JVC (UK) Ltd (Miss S Lowy; Mr K Lowy); Lightfingers (Mr Eddie Ephraums; Mr Max Ferguson); Ministry of Information (Khamis Shambah Taufik; Moosa Mahoon al Omiry; Omar Abdul Rahim; Saleh Al Ghamari); Panorama Magazine (Mr Bill Bowman; Mr Alastair Forest; Ms Michelle Barnes); Sultan Qaboos University (Mr Michael Neff); Miss Victoria Southwell; CETA Colour Laboratories.

Quartermaster
Bausch and Lomb (Mr Richard Holstock; Ms Denise McCree); Bowater Containers (Mr W J Robb; Mr John Foxton); Buffalo Bags (Mr Hamish Hamilton); Field and Trek (Equipment) Ltd (Mr I Gundle); Honda (UK) Ltd (Mr David Marsh); Karrimor International Ltd (Mr Laurie Gray); McBain Whitehouse Group (Mr Chris Naylor; Mr Everard Whitehouse); Penguin Books Ltd (Mr Len Ainsworth; Ms Maria McCabe); Philip Harris International Ltd (Oman) (Mr Richard Cochrane); Rolex Watch Company Ltd (Mr J A Nelson); Rotunda (Ms Anne Johnston; Mr A L Watson); Silva Compasses (Mr Tony Wale); Sultan of Oman's Land Forces (Raa'id D C C Doyle; MUL/1 Abdullah bin Batti Al Alawi; Raa'id Mike E Roythorne; Wkl/2 Tufail Ahmed); Supreme Plastics (Miss I Roeder); Walter Coles and Co Ltd (Mr Walter Coles); Mr John S Barker; Mr and Mrs I Jago.

Radios and Communications
Airwork Ltd (Mr Brian Little; Mr Ian MacKenzie; Mr John McInnes; Mr Michael Bragg; Mr Bob Dibben; Mr E J Mann; Mr Jeff Lee; Petroleum Development Oman (Mr David Beare); Racal Tacticom Ltd (Mr Garth Priestley; Mr Robert J Peters; Mr Mike Rose); Sultan of Oman's Land Forces Signals HQ (John Shepherd; Muqaddam A R Bailey; Naqeeb Frank Gadman; Raa'id David J Finighan; Naqeeb Mike Roberts; Raa'id Tom B Scarff; Rqb/1 Salim bin Abdullah bin Said Al Baluchi; N/Arf Hamed bin Hilal bin Mohd Al Riyami; Said Dhyab).

Transport
British Petroleum Arabian Agencies Ltd (Mr Stuart K Johnston; Mr Steve Harkin); Charles Kendall Freight Ltd (Mr Michael B Coney); Land Rover Ltd (Mr Matthew Sharp; Mr Todd Sharvell; Ms Helen Miller); Manchester Garages (Mr Bob Stoodley); Mohsin Haider Darwish (Colonel Sharma; Mohsin Haider Darwish; Mr M M Thomas; Mr Douglas N Axcell); Ramsay Ladders (Mr Gordon Lowson); Royal Society for the Prevention of Accidents (Mr M A Collins); Sultan of Oman's Air Force (Flt Lt Niall C Eden; Muqaddam (Jawwi) Peter Borrett; Muqaddam (Tayyar) John Annan; Muqaddam Rukn Tayyar Yayha bin Rasheed bin Rashid Al Juma; Naqeeb (Tayyar) Martin Hale; Naqeeb (Tayyar) Mohammed bin Khamis bin Duhai Al Busaidi; Naqeeb (Tayyar) Peter Hitchcock; Naqeeb (Tayyar) Rashid Said; Naqeeb Tayyar W D MacGillivray; Raa'id (Tayyar) Farah bin Salim bin Jabal Al Busaidi; Raa'id (Tayyar) Ted G Wood; Naqeeb (Jawwi) Ali bin Salim bin Ali Al Kilbani; Raqeeb (Jawwi) Hamood bin Yahya bin Khalfan Al Lamki; Wakeel-2 (Jawwi) Salim bin Murshid bin Said Al Farsi); Raa'id John Davis; Sultan of Oman's Land Forces Force Transport Regiment (Raa'id Chris Beal; Arf Ismail Dad Rehman Mohammad; Arf P K Rajan; Muqaddam P A H J Benton; Nb/Arf Akbar Pir Mohammad; Raa'id Salim bin Zaharan bin Salim Al Abdesalam; Wkl/1 Dick K Bray; Wkl/2 Abdul Hamid; Rqb/1 Abdul Latif; Wkl/2 Mohammad bin Bux bin Bakhtiar Al Baluchi).

H. SURVEY

BKS Surveys Ltd (Mr John D Bruntlett); Directorate of Military Survey (Major General CN Thompson; Brigadier HC Woodrow; Mr PDJ Fletcher); Electroman (Mr Barry Gordon; Mr Les Humphreys; Mr Michael Chandler); Ericcsons (Mr Neil Stewart); London School of Economics (Mrs Jane Pugh); Magnavox (Mr Colin Beatty); National Survey Authority (Aqeed Bill J F Codd; Khalfan bin Abdullah bin Ali As-Suleimani; Majid bin Zahor bin Salim Al Abdisalam; Mr D Jordan; Mr Dennis Hughes; Mr John F Miller; Mr Laurie W Gray; Mr R Alan Summerside; Mr Roger Glover; Mr TJ 'Zed' Zorichak; Saleem bin Abdallah bin Salim Al-Hashmi; Jim Butterworth; Tom Farmer); Nortech (Mr Murray Shantz; Mr Miles Taylor); North East London Polytechnic Department of Land Survey (Mr CB Burnside; Mr Bruno Canzini); Petroleum Development Oman (Mr Mike Jensen); Sultan of Oman's Artillery (Aqeed Hugh De Fontblanc; Mr Mike Morley; Naqeeb James MacIntosh; Wakeel/2 John Hinds; Yousseff Darwish); Widnell and Trollope (Mr Johnson); Wild Heerbrugg UK Ltd (Mr Brian CN Snelling); Mr Nigel Atkinson; Mr Jonathan Walton.

Finally, I would like to add my own thanks to my two editors, Anthony Lambert and Mary Remnant, for their patience and unfailing enthusiasm, despite a few sand storms on the way; and to Roderic Dutton, my deputy, whose commitment and vision has remained steadfast, and who undertook the Herculean task of co-ordinating and editing the final scientific statement.

The Project Members, Guides and Staff

Abdulhakim bin Amor bin Nasser Al Yahamadi	Shariq	Interpreter
Ali bin Abdullah bin Ha'atrush Al Wahibi	Raqah	Camp Guard
Dr Clive Agnew	University College London	Soil moisture
Dr Robert Allison	Durham University	Dune sediments
Dr Ewan Anderson	Durham University	Dew
John Augustus	Coast Security Force	Camp Steward
Major Chris Beal	Force Transport Regiment	Logistic Adviser
Heide Beal	Muscat	Capital Base Co-ordinator
Nicola Bennett-Jones	Royal Geographical Society	Project Nurse
Dorothy Bray	Durham University	Editor's Assistant
Kevin Brown	Durham University	Prosopis
Professor Denys Brunsden	King's College London	Consultant Geomorphologist
Sonya Büttiker	Basle	Invertebrates
Professor Willie Büttiker	Basle	Invertebrates
Angela Christie	Durham University	Marketing and exchange
Professor Ron Cooke	University College London	Consultant Geomorphologist
Tom Cope	Royal Botanic Gardens, Kew	Flora
Gigi Crocker	Salalah	Crafts
James Cutler	King's College London	Survey
Major Bill Davies	Sultan's Armed Forces	Liaison Officer
Captain Chris Dorman	Directorate of Military Survey	Survey
Dr Roderic Dutton	Durham University	Scientific Co-ordinator
Chris Edens	Harvard University	Lithics
Simon Ferrey	Muscat	Photography
Steve Foerander	Coast Security Force	Camp Steward
Dr Tom Gabriel	Cardiff University	Agriculture
Michael Gallagher	Oman Natural History Museum	Small mammals and birds
Dr Rita Gardner	King's College London	Aeolianite
Colonel Stephen Gilbert	Royal Engineers	Chairman, RGS Oman Committee
Ken Glennie	Shell Exploration	Geology
James Gomes	Coast Security Force	Cook
Lawrence Gomes	Coast Security Force	Cook
Vivian Gomes	Coast Security Force	Cook

Name	Institution	Role
Professor Andrew Goudie	Oxford University	Consultant Geomorphologist
John Grannan	Durham University	Computing Resources
Captain Chris Griffiths	Coast Security Force	Operations
David Hall	London Science Centre	Navigation
Charlotte Heath	Al Khaboura Project	Crafts
Corien Hoek	Muscat	Government, entrepreneurship and finance
Mike Holman	Royal Geographical Society	Field Officer
David Jones	London School of Economics	Consultant Geomorphologist
Simon Kay	University College London	Geographical Information system
Michael Keating	Royal Geographical Society	Field Officer/Photography
Khalifa bin Dhuwayin bin Hilays Al Wahibi	Mintirib	Guide and Interpreter
Colonel Khamis bin Mohammed Al Amry	Sultan's Armed Forces	Chairman, Oman Co-ordinating Committee
Khusal Khan	Field Transport Regiment	Camp Steward
William Lancaster	Stromness, Orkney	Fishing
Sophie Laurie	University College London	Plants and moisture
Ian Linn	Exeter University	Predators
Chris McBean	Durham University	Remote Sensing
Dr Judith Maizels	Aberdeen University	Palaeochannels
Mohammed Ali	Desert Regiment	Camp Steward
Dr Paul Munton	Kent University	Biological Resources Programme Director
Robert Murray-Willis	Royal Geographical Society	Navigation
Rebecca Ridley	Royal Geographical Society	Education Co-ordinator
Said bin Jabber bin Hilays Al Wahibi	Mintirib	Guide
Salim bin Hamed bin Rashid Al Hagari	Mintirib	Camp Guard
Nick Theakston	Peat Marwick Mitchell	Field Director
Richard Turpin	Royal Geographical Society	Photography
Dr Andrew Warren	University College London	Earth Sciences Programme Director
Roger Webster	Exeter University	Human Programme Director
Ernie Wickens	University College London	Survey Programme
T D R Wijesiriwardens	Coast Security Force	Cook
Shane Winser	Expedition Advisory Centre	Home Agent
Nigel Winser	Royal Geographical Society	Project Director

Bibliography

This is a selected bibliography that might be of interest to those visiting the Wahiba Sands or concerned with sand sea deserts in general.

Dr Abou Bakr Ahmed Ba Kader *et al.* (1983) *Islamic Principles for the Conservation of the Natural Environment*, (International Union for Conservation of Nature and Natural Resources – IUCN).

Anderson, E. (1986) *Wadi Al Batha*, (Public Authority for Water Resources, Muscat, May 1986).

Anderson, E. (1986) *Some Aspects of Dew Measurement*, (Public Authority for Water Resources, Muscat, May 1986).

Azzi, Robert (1973) 'Oman, land of frankincense and oil', *National Geographic Magazine* (1973) 2, pp. 205–29.

Bagnold, R. A. (1941) *The Physics of Blown Sand and Desert Dunes*, (Methuen).

Bagnold, R. A. (1951) 'Sand formations in Southern Arabia', *Geographical Journal* vol. CXVII, part I, March 1951, pp. 78–86, map.

Bertram, G. C. L. (1948) *Fisheries of Muscat and Oman*, (published by the order of the Sultan of Muscat and Oman and Dependencies).

Bloy Slade, Avril (1987) 'A Journey to Oman's Wahiba Sands', *Gulf Air News* 33, (Bahrain).

Bosch, Donald and Eloise (1982) *Seashells of Oman*, (Longman).

Brundtland, Grottarlem (1987) *Our Common Future*, (World Commission on Environment and Development).

Brunsden, D. and Cooke, R. (1987) 'The Wahiba Sands. A description of the Wahiba Sands of Oman', (Oman Ministry of Education, Muscat). In Arabic, 93 pages, March 1987.

Brunsden, D. and Cooke, R. (1987) 'The Wahiba Sands. A description of the Wahiba Sands of Oman', (Royal Geographical Society, London). Draft, in English, 75 pages, March 1987.

Carter, J. R. L. (1982) *Tribes in Oman*, (Peninsular Publishing).

Clements, Frank A. (1981) *World Bibliographical Series* vol. 29, (Clio Press, Oman).

Cooke, R. U. (1985) 'Charting the Wahiba Sand Sea', *Geographical Magazine*, vol. LVII, no. 5, May 1985, pp. 268–73.

Cooke, Ronald U. and Warren, Andrew (1973) *Geomorphology in Deserts*, (B. T. Batsford, London).

Cox, Major General Sir Percy (1925) 'Some excursions into Oman', *Geographical Journal*, vol. LXVI, no. 3, September 1925, pp. 193–227, map.

Dorman, C. (1985) 'Oman Wahiba Sands Project Route Marking and Survey Report' (unpublished).

Dutton, R. W. (1983) 'Oman', in Bowen-Jones, H. and Dutton, R. W. (eds.), *Agriculture in the Arabian Peninsula*, Economist Intelligence Unit, Special Report no. 145 (EIU, London), pp. 37–62.

Dutton, R. W. (ed.) (1986) *Oman Wahiba Sands Project Rapid Assessment Document*, (Royal Geographical Society).

Dutton, R. W., Munton, P., Warren, A., Webster, R. and Winser, N. de N. (1985) *Oman Wahiba Sands Project Mapping Phase Report* (Royal Geographical Society).

Dutton, R. W. and Winser N. de N. (1987) 'The Oman Wahiba Sands Project', *Geographical Journal*, vol. 153, no. 1, March 1987, pp. 48–58.

Edens, C. (forthcoming) *The Rub al-Khali Neolithic Revisited: The View From Nadqan, Essays in Arabian Archaeology*, (Routledge & Kegan Paul).

Gallagher, M. and Woodcock, M. W. (1980) *The Birds of Oman*, (Quartet).

Gallagher, M. and Woodcock, M. W. (1985) *Tayyur Oman*, (Quartet).

Gardner, R. A. M. (1983) 'Aeolianite' in Goudie, A. S. and Pye, K. (eds.), *Chemical Sediments in Geomorphology* (Academic Press, London).

Gibb, Sir Alexander, and Partners and W. S. Atkins and Partners Overseas, (1985) *Preliminary Soil and Groundwater Survey of Wadi al Batha in Sharqiya Region*, vol. 1, (Ministry of Agriculture and Fisheries of the Sultanate of Oman).

Glennie, K. W. (1970) *Desert Sedimentary Environments*, (Elsevier, Amsterdam).

Glennie, K. W. (1984) 'The origin of the Wahiba Sands, Oman', (MS 8 pages).

Goudie, A. S., Warren A., Jones, D. K. C. and Cooke, R. U. (1987) 'The sediment of the Wahiba Sand Sea', *Geographical Journal*, vol. 153, pp. 231–56.

Grainger, Alan (1986) *Desertification Earthscan*, (International Institute for the Environment and Development, London and Washington).

Graz, Liesl (1982) *The Omanis: Sentinels of the Gulf*, (Longman).

Groundwater Development Consultants (International) Ltd. (1982) 'Soil and groundwater survey for Kamil Wafi area', draft of final report, vol. 1: Summary, (Ministry of Agriculture and Fisheries, Oman).

Harrison, D. L. (1965) *The Mammals of Arabia Vol. I: Insectivora, Chiroptera, Primates*, (Ernest Benn).

Harrison, D. L. (1968) *The Mammals of Arabia Vol. II*, (Ernest Benn).

Harrison, D. L. *et al.* (1977) 'The scientific results of the Oman flora and fauna survey 1975', *Journal of Oman Studies* Special Report, (Ministry of Information and Culture, Sultanate of Oman).

Hawley, Donald (1977) *Oman and its Renaissance*, (Stacey International).

Holman, M. S. (1988) *Oman Wahiba Sands Project Technical and Administrative Report*, (Royal Geographical Society/Peat Marwick McLintock).

Jones, J. R. (1986) 'Results of Test Drilling for Water in Northwestern Sharqiya Area, Sultanate of Oman 1983–1984', (Public Authority for Water Resources), Report 85–21.

King, R. and Stevens, J. H. (1973) 'A bibliography of Oman 1900–1970', (Occ. papers Ser., Centre for Middle Eastern and Islamic Studies, University of Durham 2).

Lawton, J. (1988) 'Secrets of the Sands', *Aramco World*, 39, pp. 6–15.

Lawton, R. M. (1980) 'The forest potential of the Sultanate of Oman', Report to the Minister of Agriculture, Sultanate of Oman, ODA Land Resources Development Centre: Project Report 97 OMAN-03-1/REP-97/80. 39pp incl. appendices.

Lawton, R. M. (1985) 'Some indigenous economic plants of the Sultanate of Oman', *Plants for Arid Lands*, (Royal Botanic Gardens, Kew).

Larsen, Torben and Kiki (1980) *Butterflies of Oman*, (Bartholomew Books, Edinburgh).

McKee, Edwin D. (ed.) (1979) 'Global sand seas', US Geological Survey, Survey Professional Paper 1052. Includes 'The Wahiba Sands – review of previous work', [5cc], 1 page.

Maizels, J. K. (1986) *Plio-Pleistocene Raised Channel Systems of the Wahiba, Oman*, Frostick, L. E. (ed.), special publication of the Geological Society of London.

Mangin, Arthur (1869) *The Desert World*, (T. Nelson & Sons, London).

Milton, E. J. (1980) 'A portable multiband radiometer for ground data collecting in remote sensing', *Int. J. Remote Sensing* 1 (2) 153–65.

Morris, James (1957) *Sultan in Oman*, (Faber & Faber).

Munton, Paul (1984) 'Oman 1974–1984. Decade of progress. World conservation strategy in action', IUCN Bulletin Supplement 3, September 1984, pp. 4–6.

Munton, Paul (1985) 'The conservation of the Arabian Tahr Hemitragus Jayakari, IUCN/WWF Joint Op. Project 1290', *Journal of Oman Studies.*

Munton, P. N. (1985) 'The Ecology of the Arabian Tahr (Hermitragus Jayakari)', *Journal of Oman Studies*, vol. 8, pp. 11–48.

Munton P. (1987) 'Riddles of the Sands', *Geographical Magazine*, vol. 59, pp. 490–7.

Phillips, Wendell (1966) *Unknown Oman*, (Longmans).

Phillips, Wendell (1967) *Oman: A History*, (Longmans).

Ridley, R. (1988) *The Wahiba Sand Sea: A Desert Resource Pack* (BP Educational Service, PO Box 5, Wetherby, West Yorkshire, LS23 7EH).

Robinson, M. B. and Seely, M. K. (1980) 'Physical and biotic environments of the southern Namib dune ecosystem', *Journal of Arid Environments*, 3, pp. 183–203.

Seely, M. K. (1978) 'The Namib dune desert: an unusual ecosystem', *Journal of Arid Environments*, 1, pp. 117–28.

Seely, M K. and Louw, G. N. (1980) 'First approximation of the effects of rainfall on the ecology and energetics of a Namib desert dune ecosystem', *Journal of Arid Environments*, 3, pp. 25–54.

Shannon, Michael Owen (1978) *Oman and Southeastern Arabia: A Bibliographic Survey*, (G. K. Hall, Boston, Mass.).

Shaw Reade, *et al.* (1980) 'The scientific results of the Oman flora and fauna survey 1977 (Dhofar)', Office of the Government Adviser for Conservation of the Environment, Oman, *Journal of Oman Studies* Special report no. 2.

Sheppard, Tom (1984) *Desert Expeditions*, (Expedition Advisory Centre, Royal Geographical Society).

Skeet, Ian (1974) *Muscat and Oman: The End of an Era*, (Faber & Faber).

Skeet, Ian (1985) *Oman Before 1970*, (Faber & Faber).

Smythe, Kathleen (1983) *Seashell of the Sultan Qaboos Nature Reserve at Qurm. Muscat*, (Adviser for Conservation of the Environment, Muscat, Oman).

Thesiger W. T. (1950) 'Desert borderlands of Oman', *Geographical Journal*, vol. CXVI nos. 4–6, Oct–Dec 1950, pp. 137–71.

Thesiger, Wilfred (1959) *Arabian Sands*, (Longman).

Tidrick, Kathryn (1981) *Heart-beguiling Araby*, (Cambridge University Press).

UNEP (1983) 'Improvement of al Wasil Rangeland and its protection from Sand Encroachment', Consultative Group for Desert Control, fourth session, February 1983.

Ward, Philip (1987) *Travels in Oman – on the Track of the Early Explorers*, (Oleander Press Ltd.).

Warren A (1987) *Dunes in the Wahiba Sands*, (Bloomsbury Geographer).

Warren, A. and Kay, S. A. W. (1987) 'Dune networks', in *Desert Sediments: Ancient and Modern*, Frostick, L. and Reid, I. (eds.), Special Publication no. 35, (Geological Society, London), pp. 205–12.

Webster, R. M. (1985) 'Rangeland use and conservation: is the Hema system relevant? (unpublished).

Webster, R. M. (1986) 'Nomads and Development in Oman's Wahiba Sands', in *BRISMES: Proceedings of the 1986 International Conference on Middle Eastern Studies*, pp. 365–74 (British Society of Middle Eastern Studies, 1986).

Webster, R. M. (1986) 'Place names in the Wahiba Sands', (unpublished, 4 pages).

Wickens, G. E., Goodwin, J. R. and Field, D. V. (1985) 'Plants for arid lands', *Proceedings of the Kew International Conference on Economic Plants for Arid Lands, 23–27 July 1984*, (George Allen & Unwin).

Winser, N. de N. (1986) 'Wahiba Sands, The Sea of Sands and Mists', in *A Tribute to Oman*, pp. 72–9 (Oman: Apex Publishing).

Winser, Shane (1987) 'Sands yield secrets', *PDO News* 2, (Muscat, Oman).

Glossary

ABYA	large rug
AEOLIAN	windblown
ALHAMDULILAH	'God be praised'
AS SALAAM 'ALAIKUM	'Peace be with you'
'ALAIKUM AS SALAAM	'And with you, peace'
BADIYAH	area to the north of the Sands
BARCHAN	distinctive crescent-shaped sand dune
BARUSTI	date palm leaves (often used to make shelter in the Sands)
BARZAMANITE	new pink clay sediment found in the raised channels near Barzaman
BEDU	generic term for all nomadic and recently settled tribes people
BISMALLAH	'In the name of Allah'
CSF	Coast Security Force
DHIYAB	Bedu name for Roger Webster meaning 'the wolf'
DISHDASHA	traditional floor-length robe of Omani men
DURRI	large basket with a lid
FALAJ	manmade canal to bring water from the wadi to the village
FLUVIAL	water borne
GHAF	Arabic term for the *Prosopis cineraria* woodlands
GUERBA	goat-skin water container
HABL	rope

HAFLA	feast
HALAL	lawful according to the Koran
HALWA	sweet jelly-like delicacy
HARAM	forbidden
INSHALAH	'If it is the will of God'
JAMAL	male camel
JEBEL	mountain
JINNS	spirits and hobgoblins
JUNDIIS	soldiers
KHANJAR	traditional Omani silver dagger
KHARIF	monsoon wind
KHUMAR	man's embroidered hat
MA' AS SALAAMAH	'Goodbye'
MAHOWIE	woven cloth attached to a camel's saddle
MAJLIS	room set aside for receiving visitors
MASARRA	headscarf often tied in distinctive pattern
MUASKA AL MURTAFAA	Garrison headquarters of the Sultan of Oman's Armed Forces
MUDIIR	manager
MUMTAZ	excellent, magnificent
NA'IB WALI	Deputy Wali
NIMR	leopard
NIQDA	vegetated mound of sand (sometimes up to 4 metres high)
PROSOPIS	General name for the distinctive *Prosopis cineraria* tree or woodlands
QAHWA	coffee
QARHAT MU'AMMAR	site of the Field Base on the east side of the Sands
QUHAYD	site of project research base at the coast
RA'AID	military title meaning Major
RAMLAT	desert
RAS DHABDHUB	site of project research base in central Sands
RHEIM	gazelle
SABKHA	flat area, once flooded but now a salt pan
SAF	Sultan of Oman's Armed Forces
SHAMAGH	military term for headscarf
SHARQIYA	regional name meaning 'eastern region'
SHEIKH	tribal chief
SHUKRAN	'Thank you'
SUUQ	market
TAKLIIF	pain
TAMAR	dried dates
TAYLORBASE	the project's headquarters near Mintirib

TAWI	well
TAWI SARIM	site of project research base in the west side of the Sands
WADI	dry river bed
WALI	Regional Governor
WIZAR	wrap-around cloth worn by men under the *dishdasha*
YALLA	'Let's go'

Index